John Macoun

**The cryptogamic flora of Ottawa**

John Macoun

**The cryptogamic flora of Ottawa**

ISBN/EAN: 9783337271596

Printed in Europe, USA, Canada, Australia, Japan

Cover: Foto ©berggeist007 / pixelio.de

More available books at **www.hansebooks.com**

# THE CRYPTOGAMIC FLORA

# OF OTTAWA.

BY

## JOHN MACOUN, M.A., F.L.S., F.R,S.C.

Reprinted from THE OTTAWA-NATURALIST, Vol. XI, Nos. 7, 9, 11, and Vol. XII, Nos. 2, 3, 5. Issued October, 1897, to September, 1898.

# THE CRYPTOGAMIC FLORA OF OTTAWA.

By Prof. John Macoun, M.A., F.L.S., F.R.S.C.

## INTRODUCTORY NOTE.

For a number of years the writer has been collecting and studying the Cryptogamic flora of Ottawa and the surrounding country, during his holidays and in spare hours. At the present time his notes and observations have accumulated to such an extent that he considers it better to publish an incomplete list rather than wait until his excursions could take in a wider area and include a larger number of species.

Dr. James Fletcher, in "*Flora Ottawaensis*," includes an area of about 30 miles around Ottawa, and the writer purposes to 'cover about the same radius, but owing to want of time and other causes, only the district close to the city has been properly examined. It is the writer's intention to continue this work and from time to time publish the additions made by himself or others. The aim of the writer has been to place in the herbarium of the National Museum a complete set of all the species enumerated, but where possible a characteristic specimen of each species has been laid aside so that should the day ever come when the local flora of our city and its vicinity be gathered into one herbarium the Cryptogams will be forthcoming. I may then say that every species which appears in the following lists is held in duplicate and can be seen and examined at any time by those interested in the study of botany.

My notes extend over many years, as my first collections were made in the autumn of 1883, and have continued up to the present time. Owing to my absence from the city every summer, my collecting is chiefly done in September and October, and hence many fungi that are quite common around the city do not appear in the lists. Musci, Hepaticæ, and Lichens are more fully represented, but there are many species yet to be detected when my excursions become more widely extended.

The chief excuse for publishing these lists at the present time is that our club may be shown what a field for research lies at its very doors and how easily any one desirous of doing something in the botanical field can find work ready to his hand. In the following lists the dates following a locality mean the date when the specimen in our herbarium was collected.

# MUSCI.

## I. SPHAGNUM.      PEAT MOSS.

1. **S. fimbriatum,** Wilson.

Our peat bogs contain many species of Sphagnum, but none have been carefully examined except the Mer Bleue near Eastman's Springs, 12 miles from the city. This species has been gathered in the swamp on the Glebe property, Bank St. ; in the Mer Bleue, and near Casselman on the C.A.Ry.

2. **S. Girgenshonii,** Russ.

This species is chiefly found amongst black ash, growing in rigid hummocks. Collected in the swamp at the north-east corner of Beechwood Cemetery.

Var. **hygrophilum,** Warnst.

This form has been found on the Glebe property and in the swamp on the north side of Beechwood Cemetery.

3. **S. fuscum,** (Schpr.) var. **fuscescens,** Warnst.

This is a common species in all peat bogs, and is particularly abundant in the Mer Bleue ; at Casselman ; and in the swamp on the Glebe property. This is the dull rusty-coloured form.

Var. **pallescens,** Warnst.

Very common in the Mer Bleue and certainly in all large bogs in in the district.

4. **S. tenellum** (Schpr.) var. **rubellum,** (Wils.)

This form is very abundant is the Mer Bleue, and is easily distinguished from the preceding by its bright red colour.

5. **S. acutifolium,** (Ehrh.)

This is a very common species in all peat bogs, and takes many forms and colours, passing from white to purple and bright red. The common form is abundant in the swamp on the Glebe property, in Dow's Swamp at Casselman, and in the Mer Bleue.

Var. **versicolor,** Warnst.

This form is white and purple, and is very beautiful.   It is abundant in the Mer Bleue, and in the swamp on the Glebe property.

Var. **pallescens,** Warnst.

Growing generally in water, and always quite white.   Common in
the Mer Bleue.

6. **S. recurvum,** (Beauv.) var. **pulchrum,** Lindb.

This species prefers the borders of peat bogs, water-holes in them,
and black ash swamps, and takes many forms.  Its forms are recognized
by their recurved leaves.   Wet woods along the borders of the Mer
Bleue.

Var. **mucronatum,** Russ.

Wet woods along the Mer Bleue.

Var. **amblyphyllum,** Russ.

In water in holes in the Mer Bleue,

Var. **parvifolium** (Sendt.)

In the Mer Bleue and in the swamp north of Beechwood
Cemetery.

7. **S. cuspidatum,** (Ehrh.) var. **submersum,** Schpr.

Much like the preceding, but leaves not recurved.   In the swamp
on the Glebe property, Bank St.

8. **S. squarrosum,** Pers , var. **speciosum,** Russ.

This is a very beautiful species, generally found in hollows in
damp woods.   Its leaves are always very much recurved.   In damp
woods north of Beechwood Cemetery ; at Casselman ; and in woods by
the Mer Bleue.

9. **S. Wulfianum,** Girg.

Abundant in spots in the Mer Bleue.   A beautiful species.

Var. **macroclada,** Warnst.

In wet spots in the woods north of Beechwood Cemetery.

Var. **viride,** Warnst.

Swamp north of Beechwood Cemetery ; and in the swamp on the
Glebe property, Bank St.

10. **S. cymbifolium,** Ehrh.

This, the next two species and *S. acutifolium* form the bulk of
peat moss found in the bogs of Europe and America and produce the

*litter* now so extensively used iu the large cities of Europe and America. Abundant in Dow's Swamp, on the Glebe property and in the Mer Bleue.

**11. S. papillosum,** Lindb.

Abundant in the open parts of the Mer Bleue.

**12. S. medium,** Limpr. Var. **purpurascens,** Russ.

This species has been taken for a red or purplish variety of No. 10, but it is considered quite distinct. Abundant in the Mer Bleue.

### II. EPHEMERUM. Hampe.

**13. E. minutissimum,** Lindb.

Abundant on the indundated ground on both sides of the discharge from Leamy's Lake, near Hull, Que. Sept. 16th, 1889. Fruiting.

### III. ARCHIDIUM, Brid.

**14. A. ohioense,** Sulliv.

On inundated soil along the outlet of Leamy's Lake, Hull, Que. Fruiting in September.

### IV. GYMNOSTOMUM, Hedw.

**15. G. curvirostum,** Hedw.

Under wet ledges along the Ottawa at Rockliffe, near the old mill, Nov. 9th, 1896 ; also on wet rocks, Kingsmere, near Chelsea, Que. Fruiting in summer.

**16. G. rupestre,** Schw.

On wet limestone ledges at the east side of Rockcliffe, near the old mill, May 7th, 1896.

### V. WEISSIA, Hedw.

**17. W. viridula,** Brid.

On earth in woods east of Leamy's Lake, Que. ; collected on earth along the cliff, Rockcliffe Park, April 22nd, 1896. Fruiting in September.

### VI. CYNODONTIUM, Schimp.

**18. C. Wahlenbergii** (Brid.)

On dead and decaying logs in woods near Leamy's Lake, Hull, Que. ; at Meeche's Lake, north of Chelsea, Que., Sept. 23rd, 1893.

### VII. DICRANELLA, Schimp.

**19. D. varia,** Schimp.

On springy or wet clay banks. Fruiting in September. Sides of

ditches along the railway on the Experimental Farm ; also along the
Parry Sound Railway west of Hintonburg ; on the sides of the dis-
charge of Leamy's Lake, near Hull, Que. ; in a C. A. Ry. cutting at
Moose Creek.   Sept. 6th, 1889.

20. **D. heteromalla,** Schimp.

Common in sandy woods or on the roots of turned-up trees and
by roadside banks.   Fruiting in summer.   Woods at Ironsides and
Chelsea, Que. ; also at Casselman and Carleton Place ; in woods at
West End Park, Ottawa ; in McKay's Woods.   Sept. 12th, 1889.

### VIII. DICRANUM, Hedw.

21. **D. montanum,** Hedw,

On decaying logs, and stumps and bases of standing trees in
woods.   Does not fruit at Ottawa.   On the bases of trees at Leamy's
Lake, Hull, Que. ; on stumps on "Pine Hill," Rockcliffe Park, April
14th, 1896.

22. **D. fulvum,** Hook.

On boulders in woods.   Not rare in McKay's woods, but seldom
fruiting ; on boulders " Pine Hill," Rockcliffe Park ; on rocks, Aylmer
Road west of Hull, Que.   October 19th, 1891.

23. **D. viride,** Schimp.

On the bases of growing trees ; always barren.   Quite common in
McKay's Bush and Beechwood Cemetery ; in woods at Buckingham,
Que., May 14th, 1896.

24. **D. flagellare,** Hedw.

On decaying logs in damp or shady woods.   Fruiting in spring.
Meeche's Lake and Chelsea, Que. ; on " Pine Hill," Rockcliffe Park.
April 28th, 1896.

25. **D. scoparium,** Hedw.

Very common on earth in all woods around Ottawa.   Rockcliffe
Park, April 22nd, 1897.

26. **D. scopariiforme,** Kindb.

Intermediate between *D. scoparium*, Hedw. and *D. fuscescens*,
Turn. Dioecious. Leaves greenish-yellow, flexuous, lanceolate, subulate
with a short and flat subula ; margin nearly flat or slightly incurved,
densely and sharply serrate to one-third ; cell-walls rarely interrupted
by pores ; upper cells oblong-oval, lower not much narrower, inner
basal light brown ; costa thick, percurrent, with two serrate ridges
at the back in the upper part.   Capsule curved, not striate ; pedicel
red, and short.

On earth and logs in damper woods than the preceding species. Damp and wet logs in the swamp north of Beechwood Cemetery ; on rocks Meeche's Lake, near Chelsea, Que.    Sept. 23rd, 1893.

### 27.  D. fuscescens, Turn.

On old logs in Dow's Swamp ; at Chelsea and Kingsmere, Que. Fruiting on old logs near Leamy's Lake, Hull, Que. Sept. 6th, 1889.

### 28.  D. Bonjeani, DeNot.

On earth in Dow's Swamp ; in wet woods along the borders of the Mer Bleue.  Aug. 26th, 1889.  Barren.

### 29.  D. undulatum, Turn.

Common in cool damp woods on earth.  Stewart's Bush, Dow's Swamp, Mer Bleue, and McKay's Woods ; on the cliffs, Rockcliffe Park, April 22nd, 1896.  Fruiting in summer.

### 30.  D. spurium, Hedw.

On Laurentian rocks on Gilmour's Island, Chelsea, Que.  May 22nd, 1892.  Barren.

### IX.  FISSIDENS, Hedw.

### 31.  F. bryoides, Hedw.

On earth in woods between St. Patrick's Bridge and Beechwood Cemetery, east of the road ; on earth in woods Leamy's Lake, Hull, Que. Oct. 16th, 1889.  Fruiting.

### 32.  F. minutulus, Sulliv.

On stones in the channel of the small brook entering McKay's Lake near Beechwood Cemetery, Oct. 12th, 1884.  Fruiting.

### 33.  F. pusillus, Wils.

Abundant on damp, flat, limestone rocks in McKay's Woods, south-west of the lake.  Oct. 12th, 1884.  Fruiting.

### 34.  F. osmundoides, Hedw.

On earth on turned-up trees in Dow's Swamp ; on roots of trees in woods at Leamy's Lake, Hull, Que. ; on roots of trees in old woods at Carleton Place.  May 31st, 1884.

### 35.  F. decipiens, DeNot.

Very abundant on turned-up roots and old stumps in Dow's Swamp ; on earth in woods at Leamy's Lake, Hull, Que. ; also in McKay's Bush near the lake ; collected on damp rocks, Rockcliffe Park, April 22nd, 1896.

## X. LEUCOBRYUM, Hampe.

### 36. L. vulgare, Hampe.

On earth in damp woods north of Beechwood Cemetery : also in woods on " Long Point," Mer Bleue ; on the banks of the Lievre River at Buckingham, Que, May 14th, 1896. Seldom fruiting.

## XI. CERATODON, Brid,

### 37. C. purpureus, Brid.

Very common everywhere in pasture fields, by roadsides, on old fences and roofs of houses. Our commonest moss, and found in all parts of the habitable earth. Fruiting in early spring. With mature fruit, May 12th, 1896.

## XII. SELIGERIA, Bruch and Schimp.

### 38. S. campylopoda, Kindb.

Agrees with *Seligeria recurvata* in the shape of the capsule and the arcuate pediel, but differs considerably in the leaves being broader, very much shorter, sublinear, obtuse, rarely short-acuminate and sub-acute, and the costa not excurrent, the perichetial leaves ovate-oblong, thin-costate, the peristome darker red. The male flower is fixed on the side of the female.

Under damp overhanging limestone rocks near the upper part of the Beaver Meadow, on the east side, west of Hull. Que. April 26th, 1891. Fruit nearly full grown.

### 39. S. recurvata, Bruch. and Schimp.

On large boulders by the roadside leading from the end of the Electric Railway eastward towards the old mill, Rockcliffe Park. May 7th, 1896. Fruit ripe.

## XIII. DIDYMODON, Hedw.

### D. rubellus, Bruch. and Schimp.

On damp limestone ledges near McKay's Lake ; also on ledges at Leamy's Lake, Hull ; Chelsea and Meeche's Lake, Que. ; on damp limestone rocks Rockcliffe Park. April 22nd, 1896. Fruiting.

## XIV. LEPTOTRICHUM, Hampe.

### 41. L. tortile, C. Muell.

Roadside near the Mer Bleue ; at Eastman's Springs, Sept. 29th, 1892 ; on an old road in woods at the end of the Electric Road, Rockcliffe Park.

**42. L. glaucescens,** Hampe,

On calcareous earth in crevices of rocks along lakes and rivers. Along the outlet of Leamy's Lake, south side ; at Kirk's Ferry and Meeche's Lake ; on the cliffs facing the Ottawa, Rockcliffe Park. April 22nd, 1896.

## XV. BARBULA, Hedw.

**43. B. brevirostris,** Bruch. and Schimp. (?)

On large boulders, growing with *Seligeria recurvata* along the road leading east from the end of the Electric Road at Rockcliffe Park, May 7th, 1896. Fruiting.

**44. B. tortuosa,** Web. and Mohr.

On rocks near McKay's Lake and around the cliffs, Rockcliffe Park ; on " Pine Hill," Rockcliffe Park, April 16th, 1896 ; on rocks Meeche's Lake near Chelsea, Que., Sept. 23rd, 1893.

**45. B. unguiculata,** Hedw.

Very common, some years, on old roads and streets in and around Ottawa. Mackenzie Ave., Oct. 12th 1896 ; on limestone rocks by the Ottawa, Rockcliffe Park.

**46. B. convoluta,** Hedw,

Quite common in pastures, growing with *Ceratodon purpureus*, known by its *yellow* pedicels. By roadsides and in pastures at the Experimental Farm and north-west to Hintonburg ; also by the C.P.Ry. at Carleton Place ; on earth in pastures at Rockcliffe Park, May 12th, 1896. Fruiting early in spring.

**47. B. ruralis,** Hedw.

Generally found on limestone shingle or gravelly ridges. Rockcliffe Park near Governor's Bay ; at Britannia and along the railway at Carleton Place. Barren.

## XVI. GRIMMIA.

**48. G. apocarpa,** Hedw.

On boulders everywhere around Ottawa ; especially in McKay's Woods ; at Meeche's Lake and Chelsea, Que. ; at Carleton Place, and Stittsville ; on " Pine Hill," Rockcliffe Park, April 14th, 1896. Fruiting abundantly late in autumn.

## XVII. HEDWIGIA, Ehrh.

**49. H. ciliata,** Ehrh.

Quite common on boulders, McKay's Woods, and other places

around Ottawa ; on boulders " Pine Hill," Rockcliffe Park, April 16th, 1896. Fruiting.

Var. **viridis**, Schimp.

On boulders in shady woods quite common at Ottawa ; Oct. 12th, 1884.

Var. **subnuda**, Kindb.

Leaves nearly hairless, the greater number broadly ovate, borders reflexed ; cells larger, subquadrate.
On boulders in McKay's Woods near the lake, April 28th, 1896. Fruiting.

### XVIII. ULOTA, Mohr.

50. **U. Ludwigii**, Brid.

On trees along the creek in Beaver Meadow north of the toll-gate on the Aylmer Road ; very rare on " Pine Hill," Rockcliffe Park. May 7th, 1896.

51. **U. crispa**, Brid.

On cedar trees in Dow's Swamp ; and on spruce trees along the Beaver Meadow Creek west of Hull, Que. ; on spruce trees in Rockcliffe Park near Governor's Bay, April 22nd, 1896.

52. **U. camptopoda**, Kindb.

Stem not creeping. Leaves, when dry crisped, when moist patent, or squarrose, often curved, faintly papillose, from a short dilated ventricose base, suddenly narrowed into the acute or subulate acumen, borders recurved at the base, and also often above on one side ; outer basal cells, disposed in 2-5 rows, quadrate-rectangular thick-walled ; inner narrow, orange, upper rotundate ; costa elevate, stout percurrent. Capsule small, long-necked, when dry faintly plicate, narrow, subcylindric and not constricted below the mouth, obovate when moist ; teeth bigerminate, pale, when dry recurved ; cilia none ; lid long-apiculate ; pedicel long, but not much emergent, flexible, more or less curved or geniculate, in young as well as in the dry state ; calyptra densely hairy, covering the capsule.
Habit of *U. crispula.* Agrees with *U. maritima* in the curved pedicel ; differs from *U. Ludwigii* in the narrower capsule. Growing together with *U. Ludwigii* on trees along the Beaver Meadow Creek west of Hull, Que. ; also on the pales on the south-west corner of the Cemetery west of Hull on the Aylmer Road, Que. ; April 26th, 1891.

53. **U. connectens**, Kindb.

Monœcious. Tufts soft, pulvinate, green above ; blackish below. Stems erect. Leaves, from an ovate concave base, linear-lanceolate,

when dry very much crisped, when moist subarcuate, short attenuate
to the acute apex ; borders revolute above the base, for the greater
part, at least on one side, distinctly papillose, also at the back ; cells at
basal wings sub-quadrate hyaline with incrassate transverse walls, those
next the costa narrower, rectangular, in straight rows, the lowest orange ;
costa pale, sub-percurrent. Male flower at the side of the female. Inner
perigonial leaves broad, short-ovate, obtusate or suddenly short-acuminate ; cells round only in the acumen, the others narrow, the lower
basal wider and yellow ; antheridia about 9, with several paraplyses,
Perichetial leaves with sublinear basal cells.   Capsule dark-brown short
subovoid, not contracted at the mouth, costate ; pedicel short, scarcely
emergent.   Calyptra densely hairy.

This species is a true *Ulota*, although the revolute leaf-borders, the
distinctly papillose cells and short pedicellate capsule are more like an
*Orthotrichum.*

On cedar trees (*Thuya occidentalis*) in Dow's Swamp, September
16th, 1886.

Both the preceding species are believed to be forms of *U. crispa*
by Mrs. E. G. Britton, who has made a special study of the genus.

### XIX. ORTHOTRICHUM, Hedw.

**54. O. anomalum, Hedw.**

On rocks and ledges along the Ottawa at Governor's Bay, Rockcliffe Park ; also on ledges near McKay's Lake, in fruit April 22nd,
1896.   Fruiting.

**55. O. speciosum, Nees.**

Common on balsam fir, cedar and spruce trees in the woods east
of Beaver Meadow west of Hull, Que. ; also on spruce trees in Rockcliffe Park ; collected on trees and fence rails near Hintonburg, April
13th, 1896.   Fruiting.

**56. O. sordidum, Sulliv. and Lesq.**

Common on beech trees in woods near Ironsides, Que. ; collected
on trees in Rockcliffe Park and Beechwood Cemetery, April 22nd,
1896.   Fruiting.

**57. O Ohioense, Sulliv. and Lesq.**

On trunks in woods near Leamy's Lake, Hull, Que. ; old fence rails
at Carleton Place ; collected on trees in woods near Governor's Bay,
Rockcliffe Park, April 22nd, 1896.   Fruiting.

**58. O. Canadense, Bruch and Schimp.**

This species appears in Part VI under *O. Schimperi* but was dis-
covered by Mrs. E. G. Britton when monographing the genus some

years since. It is apparently very rare as its occurrence in America was doubted when Lesq. and James' work on the mosses appeared in 1884.

On rocks at the corner of Rockcliffe Park close to Governor's Bay. October 12th, 1884.

### 59. O. cupulatum, Hoffm.

On limestone rocks along the cliffs facing the Ottawa near Governor's Bay, Rockcliffe Park. April 16th, 1891.

### 60. O. strangulatum, Beauv.

On trunks and fences around Ottawa ; woods at Ironsides, Chelsea and near Leamy's Lake, Que. ; also in McKay's Woods and in Beechwood Cemetery ; collected on trees in Rockcliffe Park, April 22nd, 1896.

### 61. O. psilothecium, C. M. and Kindb.

Plants small, 1 cm. long or less, green Leaves short oblong-lanceolate, obtusate or short-acuminate, sub-obtuse, revolute at the borders to the greater part, faintly papillose ; costa percurrent, Capsule small, immersed, oblong, not striate before sporosis, very short-necked ; vaginula naked ; calyptra slightly hairy at the blackish apex, finally glabrous and light-brown, narrow, covering the whole capsule ; lid rostellate. Male flowers on distinct branches.

This species has the habit of *O. fallax*, Schimp. (*O. Schimperi*, Hamm.) We have not been able to examine the peristome and the stomata of the capsule, because only one capsule (in our specimen) is nearly ripe, the others are quite unripe.

On old fences in Rockcliffe Park ; on cedar rails along the Richmond Road, near Hintonburg ; collected on old fences at Carleton Place, Aug. 26th, 1889.

### 62. O. obtusifolium, Schrad.

On old cedar rails and trunks of balsam poplar ; on rails in McKay's Bush ; on poplar trees along the Gatineau River, near Leamy's Lake, Hull, Que. ; collected on poplar trees near Hintonburg, April 18th 1896.

## XX. ENCALYPTA, Schreb.

### 63. E. vulgaris, Hedw.

On limestone ledges on the south side of the outlet of Leamy's Lake, near the Hull Cemetery, Que., Oct. 11th, 1890.

### 64. E. Macounii, Austin.

In crevices of limestone rocks around the whole cliff facing the Ottawa in Rockcliffe Park, April 22nd, 1896 ; crevices of rocks along the Gatineau at Kirk's Ferry, Que. Fruiting.

## 65. E. streptocarpa, Hedw.

On limestone rocks at the outlet of Leamy's Lake, near Hull
Cemetery, Que., Sept. 6th, 1889 ; on the cliffs at Governor's Bay,
Rockcliffe Park.   Barren.

### XXI. TETRAPHIS, Hedw.

## 66. T. pellucida, Hedw.

On the bases of stumps and dead logs (chiefly pine and cedar), in
all swamps and wet woods around Ottawa   On old stumps in Dow's
Swamp, and on Cowley's Farm, near Hintonburg ; collected April 18th,
1896.   Fruiting.

### XXII. PHYSCOMITRIUM, Brid.

## 67. P. immersum, Sulliv.

On inundated alluvial soil (in small tufts) along the outlet of
Leamy's Lake, near Hull, Que. Sept. 16th, 1889.   Fruiting.

## 68. P. platyphyllum, Kindb.

Lower leaves sublingulate, yellow-margined, serrate all round, with
a percurrent costa ; the upper very broad, ovate-acuminate ; indistinctly
margined, serrate above the middle, costa percurrent or short excurrent ;
cells wide sub-hexagonal, the basal sub-rectangular, all hyaline. Calyptra
mitriform.    Capsule pyriforme ; lid mammillate ; pedicel (unripe)
yellow, about 1 cm. long, or shorter.

Since this description was published Mrs. E. G. Britton has
examined the specimen and pronounces it *P. turbinatum*, Muell.
Better specimens are wanted to settle the question, but houses and
lawns and asphalt cover where it was found by Dr. Fletcher many
years ago.

On earth at the southern end of Metcalfe Street, Ottawa.

### XXIII. FUNARIA, Schreb.

## 69. F. hygrometrica, Sibth.

Very common on old walls and especially on burnt soil in damp
woods and on old turned-up roots in swamps.   Common around
Ottawa and at Carleton Place.

### XXIV. BARTRAMIA, Hedw.

## 70. B. Œderiana, Swartz.

On rocks east of the Beaver Meadow, west of Hull ; on damp
rocks, Chelsea and Kingsmere, Que. ; on limestone rocks near McKay's
Lake ; collected on the cliffs by the Ottawa, Rockcliffe Park, April
22nd, 1896.   Fruiting.

71. **B. pomiformis,** Hedw.

Crevices of damp and dripping rocks near Gilmour's Mill, Chelsea, Que. ; collected on damp limestone ledges on the cliffs facing Gatineau Point, Rockcliffe Park, April 22nd, 1896.   Fruiting.

## XXV. PHILONOTIS, Brid.

72. **P. fontana,** Brid.

By springs at Kingsmere, and Kirk's Ferry, Que.

## XXVI. LEPTOBRYUM, Schimp.

73. **L. pyriforme,** Schimp.

Rather common on burnt soil in swamps and along ditches.   Casselman and Carleton Place ; border of Dow's Swamp, Oct. 12th, 1884.

## XXVII. WEBERA, Hedw.

74. **W. nutans,** Hedw.

On rotten logs and stumps in swamps and wet woods ; common, Dow's Swamp, Kingsmere and Casselman ; common in McKay's Bush ; collected on old logs in Beechwood Cemetery, May 12th, 1896. Fruiting.   Mer Bleue June 15th, 1892.

75. **W. albicans,** Schimp.

On wet limestone rocks. under the cliffs at the end of the Electric Road, Rockcliffe Park.   Nov. 9th, 1896.

## XXVIII. BRYUM, Dill.

76. **B. pendulum,** Schimp.

On wet earth at Ottawa ; woods north of Beechwood Cemetery, October 20th, 1884.

77. **B. bimum,** Schreb.

Common in wet woods and swamps. Borders of the Mer Bleue, June 15th, 1892.

78. **B. intermedium,** Brid.

Crevices of damp rocks and old walls. In the old quarry in Rockcliffe Park and along the cliffs by the Ottawa. May 12th, 1896.

79. **B. argenteum,** Linn.

Very common on roadsides and on desiccated soil in old pastures and waste places. On earth along St. Louis Dam, October 24th, 1884.

80. **B. cæspiticium,** Linn.

On earth in pasture fields and open thickets ; open places in Rockcliffe Park, May 7th, 1896 ; in old pastures near Hintonburg, Oct. 4th, 1884.

81. **B. capillare,** Linn. Var. **heteroneuron,** C. M. & Kindb.

On roots of trees in McKay's Bush near the lake, May 24th, 1888.

82. **B. Duvalii,** Voit.

In ditches and on wet rocks. In a springy place at the end of the Electric Railway, Rockcliffe Park.

83. **B. pseudo-triquetrum,** Schwæger.

In wet woods and swamps. Swamp north of Beechwood Cemetery ; also in Dow's Swamp.

84. **B. Ontariense** Kindb.

Intermediate between *B. roseum* and *B. Beyrichii* (Hsch.), C. Mueller. Comal leaves very numerous ; lingulate, abruptly and short acuminate, revolute to ⅔ or ¾, yellow-margined above with great confluent teeth ; costa stout, excurrent. Capsule pale, with a distinct, curved collum half as long, teeth papillose and hyaline above ; archegonia numerous ; lid convex short-apiculate, not oblique.

Hitherto confounded with *B. roseum*, and quite common throughout Ontario ; generally in a barren state. On old logs and sometimes on limestone rocks in maple woods around Ottawa. Beechwood Cemetery, Rockcliffe Park, Carleton Place and Eastman's Springs ; on logs in Dow's Swamp, October 10th, 1889.

85. **B. Laweri,** Ren. and Cardt.

On rocks opposite the island in the Gatineau River, Gilmour's Park, Chelsea Que., Sept. 9th, 1889.

## XXIX. MNIUM, Linn.

**86. M. cuspidatum, Hedw.**

Quite common on earth at the roots of trees in dry woods. On earth in woods Patterson's Creek, Stewart's Bush, Carleton Place, and Beechwood Cemetery ; on earth in Rockcliffe Park, April 28th, 1896.

**87. M. rostratum, Schwægr.**

On a large boulder on "Pine Hill," Rockcliffe Park, April 16th, 1896. Barren.

**88. M. Drummondi, Br. and Sch.**

In damp or swampy woods, near High Rock, Lièvre River, above Buckingham, Que., May 19th, 1884.

**89. M. affine, Bland.**

On earth in swamps and along brooks. On roots of trees along the brook west of West End Park, October 10th, 1884.

**90 M. rugicum, Laur.**

Rather common, growing in the wettest part of Dow's Swamp. Sept. 16th, 1889.

**91. M orthorrhynchum, Br. and Sch.**

On damp limestone rocks near McKay's Lake, Aug. 26th and Oct. 12th, 1889.

**92 M. pseudo-lycopodiodes, C. Muell.**

On the bases of trees in cedar and black ash swamps. In Dow's Swamp and at Casselman ; in the swamp north of Beechwood Cemetery, May 7th, 1896. Fruiting ; on rocks in a brook, Meeche's Lake, near Chelsea, Que., Sept. 23rd, 1893.

**93. M. inclinatum, Lindb.**

On old stumps in Dow's Swamp ; on damp limestone rocks along McKay's Lake, April 22nd, 1896 ; old fruit. Old stumps in Dow's Swamp, Sept. 16th, 1889.

**94. M. spinulosum, Br. and Sch.**

On earth at the bases of trees, chiefly hemlocks. Wet woods north of Beechwood Cemetery ; woods near Carleton Place ; on the bank of the Lièvre River at Buckingham, Que., May 14th, 1896. Fruiting.

**95. M. stellare, Hedw.**

On old stumps in cedar swamps. In Dow's Swamp, May 2nd, 1896. Old fruit.

**96. M. punctatum,** Hedw.

On earth in cedar swamps and along small brooks in woods. In Dow's Swamp ; also by a brook near Meeche's Lake, north of Chelsea, Que., Sept. 23rd, 1893.

### XXX. AULACOMNIUM, Schwægr.

**97. A. palustre,** Schw.

Very common in swamps.    Mer Bleue and at Casselman.

### XXXI. TIMMIA, Hedw.

**98. T. megapolitina,** Hedw.

On roots of trees by brooks in wet woods and on wet rocks. On roots of trees in Dow's Swamp, May 2nd, 1896, by a brook west of West End Park ; and on wet rocks south end of McKay's Lake, Oct. 16th, 1884.

### XXXII. ATRICHUM, Beauv.

**99. A. undulatum,** Beauv.

On damp sandy earth in cool woods and wet sandy pastures. Common in McKay's Woods, Beechwood Cemetery, Mer Bleue and Casselman ; woods rear of Cowley's Farm, Hintonburg, April 18th, 1896.    Fruit old.

### XXXIII. POGONATUM, Beauv.

**100. P. brevicaule,** Beauv.

Along a ditch cut along the road leading through West Casselman. May 12th, 1891.

**101. P. alpinum,** Rœhl.

On damp sandy slopes and amongst rocks east side of Rideau River, near Hog's Back, May 7th, 1897.

### XXXIV. POLYTRICHUM, Linn.

**102. P. gracile,** Menz.

Not uncommon in the Mer Bleue, near Eastman's Springs, June 15th, 1891.

**103. P. Ohioense,** Ren. and Card.

On earth in woods near Casselman, east of the Nation River. May 16th, 1891.

**104. P. juniperinum, Willd.**

In old pastures and on old pine stumps in fields and by roadsides ; quite common. At the base of stumps north of Beechwood Cemetery, May 12th 1896.

**105. P. commune, Linn.   Var. Canadense, Kindb.**

Differs principally in the low stem, about 6 8 cm. high, the pedicel not much longer, the blackish capsule much shorter than in the type which probably is very rare in Canada and only once examined by Kindberg.

In wet places at Britannia, Oct. 11th, 1892 ; also in hollows in the pine woods by the lake, east of Aylmer, Que., Sept. 11th, 1892 ; wet pastures at Casselman, June 12th, 1892.

### XXXV. BUXBAUMIA, Hall.

**106. B. aphylla, Linn.**

On rocks at Chelsea, eight miles from Ottawa, Que.   (Dr. Fletcher.)

### XXXVI. FONTINALIS, Dill.

**107. F. antipyretica, Linn.**

On stones in the brook running through the Beaver Meadow west of Hull.

### XXXVIII. DICHELYMA, Myrin.

**108. D. pallescens, Bruch and Schimp.**

On the bases of trees and twigs in water-holes near St. Patrick's Bridge ; also very abundantly in the woods subject to spring floods around Leamy's Lake, Hull, Que., Nov. 9th, 1896.   Fruiting.

### XXXIX. NECKERA, Hedw.

**109. N, pennata, Hedw.**

Quite common on trunks in swamps or wet woods around Ottawa. On trunks north of Beechwood Cemetery, also in Dow's Swamp : on trunks " Pine Hill," Rockcliffe Park, April 16th, 1896.

**110. N. oligocarpa, Bruch and Schimp.**

On ledges of limestone rocks opposite Gatineau Point, Rockcliffe Park, May 7th, 1896.

### XL. LEUCODON, Schwægr.

**111. L. sciuroides, Schwægr.**

Common on trunks in all old woods.   Dow's Swamp, Carleton Place, Eastman's Springs and Casselman ; in McKay's Woods, May 7th, 1896.   Barren.

## XLI. HOMALIA, Brid.

**112. H. Macounii,** C. M. and Kindb.

*II. trichomanoides,* Lesq. and James Mosses of N. America, 245.

Very nearly allied to *Homalia trichomanoides* ; differs in the leaves being longer, rather lingulate, the lowest basal cells yellow, the perichetial leaves more suddenly narrowed to the very short acumen, the segments of the peristome cleft between the articulations.

On the bases of trees on the south end of Cowley's Farm, west of Hintonburg ; on limestone rocks east end of McKay's Lake ; on the cliffs facing Gatineau Point, Rockcliffe Park, May 7th, 1896.

## XLII. MYURELLA, Bruch and Schimp.

**113. M. julacea,** Bruch and Schimp.

On old stumps in cedar swamps. In the swamp north of the Experimental Farm ; also in Dow's Swamp.

**114. M. Careyana,** Sulliv.

On ledges of limestone rocks, east of the creek in the Beaver Meadow west of Hull, Que. ; crevices of wet limestone rocks at the north end of Rockcliffe Park near the old mill, May 7th, 1896 ; on rocks at Meeche's Lake, Sept. 23rd, 1893.

## XLIII. LESKEA, Hedw.

**115. L. polycarpa,** Ehrh.

Very abundant on the bases of trees in the inundated flats between Leamy's Lake and the mouth of the Gatineau River. Nov. 9th, 1896.

**116. L. nervosa,** Myrin.

On trunks in McKay's woods ; on boulders at the east end of Rockcliffe Park, near the old saw mill, May 7th, 1896 ; also at Carleton Place.

## XLIV. ANOMODON, Hook. and Tayl.

**117. A. rostratus,** Schimp.

Very common on the roots of trees in swamps and on the faces of perpendicular, damp rocks. Seldom fruiting. On rocks in Rockcliffe Park, April 22nd, 1896.

**118. A. attenuatus,** Heuben.

Quite common on the bases of trees in black ash swamps and wet woods. Fruiting abundantly in the autumn. On trees in Rockcliffe Park, April 16th, 1896.

**119. A. obtusifolius,** Br. and Sch.

Abundant on the bases of trees in black ash swamps and wet woods around Ottawa. Fruiting in the autumn. On trunks south end of Cowley's Farm, near Hintonburg, April 18th, 1896.

**120. A. apiculatus,** Bruch and Schimp.

On decaying logs and flat limestone rocks in McKay's Woods, opposite the entrance to Beechwood Cemetery ; on limestone rocks, Rockcliffe Park, April 22nd, 1896.

**121. A. viticulosus,** Hook. and Tayl.

Very common on limestone rocks near McKay's Lake and along the Ottawa west of Hull, and on damp ledges in woods along the Beaver Meadow west of Hull, Que. ; on damp rocks facing the river, Rockcliffe Park, April 22nd, 1896.

**122. A. heteroideus,** Kindb.

Plants densely tufted, green, finally fuscescent or blackish. Stem creeping, subpinnate, much branching and furnished with numerous small, flagelliform branchlets, densely beset with very small, oblong, obtuse and nerveless leaves ; paraphyllia broad. Stem-leaves subdistant, decurrent, appressed when dry, open-erect when moist, from a broadly ovate base suddenly narrowed to a long, subulate or sublinear acumen, entire, faintly papillose ; margins revolute at the base ; branch-leaves more attenuate ; cells round-oval, the marginal of the base quadrate ; costa vanishing below the acumen. Diœcious. Fruiting specimens not found. This species resembles *Leskea nervosa* in habit.

On flat limestone rocks and the roots of trees in McKay's Woods, opposite the entrance to Beechwood Cemetery ; on limestone rocks at Meeche's Lake, near Chelsea, Que., Sept. 23rd, 1893.

XLV. PLATYGYRIUM, Bruch and Schimp.

**123. P. repens,** Bruch  and Schimp.

On old logs in woods at Eastman's Springs, Casselman and Carleton Place ; on old logs in woods north of Beechwood, May 7th, 1896.

Var. **orthoclados,** Kindb.

Branches elongate and not curved. All basal leave-cells orange. Segments linear, not completely free at base, smooth or denticulate at one side, not shorter than the teeth.

On old logs at the south end of Skead's Farm, west of Hintonburg, May 15th, 1885.

20 THE OTTAWA NATURALIST. [December

XLVI. PYLAISIA, Bruch and Schimp.

**124. P. polyantha,** Bruch and Schimp.

On rotten logs at Carleton Place ; on trees in McKay's Woods, Oct. 12th, 1884.

**125. P. Selwynii,** Kindb. Ott. Nat. II., 156.

Differs from *P. intricata* in the denser, darker green tufts, the leaves broader, short-acuminate, reflexed to the acumen at one border or at both, the short alar and marginal cells more numerous, the capsule short-oval, the segments adhering to two-thirds of the teeth.

Very abundant on old cedar fences along the Richmond Road, three miles west of Ottawa ; also on fences west of Hintonburg, April 18th, 1896. Mr. A. J. Grout reduces this species to *P. intricata*, and believes it to be merely a more compact form.

**126. P. intricata,** Bruch and Schimp.

Common on logs and trunks, in woods and on old cedar rails around Ottawa ; on fence rails and trunks and branches in Rockcliffe Park, April 28th, 1896.

**127. P. velutina,** Bruch and Schimp.

On old logs in Dow's Swamp, Sept. 25th, 1889.

XLVII. ENTODON, C. Mueller.

**128. E. acicularis,** C. M. and Kindb.

Tufts compact, brown-yellow or variegate with green. Stems much divided, very radiculose ; branches very short and turgid, not attenuate Leaves imbricate, with difficulty loosed from the stem, scarcely open when moist, finally golden-yellow, from the ovate-oblong base suddenly narrowed to a fine aciculiform or subulate point, denticulate nearly all around ; cells not chlorophyllose, linear-lanceolate or fusiform, the alar not well defined ; costa generally wanting. Barren.

On limestone rocks in woods near McKay's Lake, May 2nd, 1885 ; also by an old lime-kiln at Britannia, Oct. 11th, 1890,

In a late revision of the genus Mr. A. J. Grout has reduced this species also to a form of the next.

**129. E. cladorrhizans,** (Hedw.)

On old logs in Rockcliffe Park and McKay's Woods, April 16th. 1896 ; on stones and logs at Britannia ; and very abundant on old logs in woods at Carleton Place. Fruiting abundantly in autumn.

XLVIII. CLIMACIUM, Web. and Mohr.

**130. C. Americanum,** Brid.

On the ground in swamps about wet woods around Ottawa. Seldom fruiting, but frequent in woods along the Beaver Meadow, Hull-

Que. ; on earth in the swamp north of Beechwood Cemetery, May
1885.

### 131. C. dendroides, Web. and Mohr.

In a swamps on the east side of the Beaver Meadow west of Hull,
near the north end, Oct. 24th, 1891.   Fruiting abundantly.

#### XLIX. PTEROGONIUM, Swartz.

### 132. P. brachypterum, Mitten.

On a small maple trunk in a piece of woods along the west side of
McKay's Lake.  Fruiting.   April 28th, 1896.   This determination is
doubtful.

#### L. THUIDIUM, Schimp.

### 133. T. minutulum, Bruch and Schimp.

On old logs in McKay's Woods ;  also in woods at Ironsides ;  on
stumps and rocks, Rockcliffe Park, April 22nd, 1896.

### 134. T. scitum, Aust.

On beech trunks in McKay's Woods ; on beech trees on the south
end of Cowley's Farm, west of Hintonburg, April 18th, 1896.

### 135. T. gracile, Bruch and Schimp.

On old logs in woods at King's Mountain, near Chelsea, Que. ;
abundant on rotten logs at Leamy's Lake, near Hull, Que., Nov. 9th,
1896.   Fruiting.

### 136. T. recognitum, (Hedw.) Lindb.

On old logs around the Mer Bleue ; in Dow's Swamp and in a
swamp at Stittsville ; on rotten wood, "Pine Hill," Rockcliffe Park,
April 16th, 1896.

### 137. T. delicatulum, Mitt.

On earth in the Mer Bleue and in the swamp north of Beechwood
Cemetery ; on earth in Dow's Swamp, May 2nd, 1896.

### 138. T. abietinum, Bruch. and Schimp.

Quite common on exposed limestone rocks at Carleton Place ; also
abundant around the cliffs of Rockcliffe Park.   April 22nd, 1896.
Always barren.

### 139. T. Blandovii, Bruch and Schimp.

In damp woods at Britannia ; in the Mer Bleue ; also on earth in
Dow's Swamp.

## LI. CAMPTOTHECIUM, Schimp.

**140. C. nitens,** Schimp.

Abundant in the Mer Bleue ; also in Dow's Swamp, June 4th, 1884.

## LII. BRACHYTHECIUM, Schimp.

**141. B. laetum,** Brid.

A common species on stones in woods around Ottawa ; on boulders in woods at the south end of Cowley's Farm, west of Hinton-burg, April 18th, 1896. Fruiting.

**142. B. digastrum,** C. M. and Kindb.

Tufts largely cohering, olive-green, not shining. Stem rigid, sub-pinnate or irregularly branching, radiculose below ; branches sub julaceus, obtusate. Stem-leaves when dry loosely appressed or sub-imbricate, crowded, patent or subsecund when moist, decurrent, not auricled, plicate, biventrose, ovate and short-acuminate with the acumen flexuous or when dry serpentino-corrugate, borders more or less re-curved but not reflxed, subentire or faintly denticulate above ; lower basal cells wide and sub-rhombic, the alar rather quadrate-rectangular and not very distinct, the upper conflate, small, very chlorophyllose, the inner median sublinear, the others oblong-lanceolate ; costa thick and subflexuous, long and vanishing near the acumen. Branch leaves ovate-oblong, more distinctly revolute at the borders, denticulate at the acumen and narrower areolate. Female flowers small, inner perichetial leaves filiform-acuminate with the acumen arcuate, long-costate and denticulate. Capsule asymmetric sub-cylindric, curved ; lid long-conic ; pedicel smooth 1-2 cm. long. Peristomial teeth conic-connivent when moist, dark red-brown below, very much longer than in the middle open segments ; cilia nodulose not apendiculate, annulus none. Monœcious.. Habit of *Leucodon julaceus.*

On rocks in McKay's Woods near south end of the lake, Oct. 12th, 1889. Fruiting.

**143. B. acuminatum,** (Beauv.)

On earth in McKay's Woods between the old entrance to Beech-wood Cemetery and the lake ; also on logs at Carleton Place.

**144. B. salebrosum,** Bruch and Schimp.

On stones in damp woods north of Beechwood Cemetery, Oct. 16th, 1884 ; on stones at the rear of Cowley's Farm, west of Hinton-burg, April 18th, 1896.

**145. B. acutum,** (Mitt.) Sulliv.

On earth in wet woods north of Beechwood Cemetery ; also in damp woods along the Beaver Meadow west of Hull, Que., Oct. 12th, 1891.

146. **B. platycladum,** C. M. and Kindb.

Tufts densely cohering, bright green, shining. Stem irregularly branching ; branches short, obtuse, complanate. Leaves loosely imbricate or patent, nearly flat, long-decurrent, distinctly auriculate, faintly striate, broad, ovate, suddenly and generally short-acuminate ; borders not recurved, faintly sinuolate or sub-entire below the middle, more distinctly denticulate above ; cells pale, the upper narrow, the lower near the base dilated, the alar large and well defined ; costa short, reaching little above the middle. Capsule sub-oval, faintly curved ; teeth dark-yellow, entire at the borders ; cilia not appendiculate ; lid unknown ; pediel rough, about 2 cm. long, or shorter. Perichetial leaves long, fiiliform-acuminate, the point arcuate. Diœcious.

Differs from *B. rutabulum* principally in the long-decurrent auricled leaves and the diœcious inflorescense ; from *B. rivulare* in the peristome, etc.

On stones in the bed and along the sides of a brook, north of the Parry Sound Railway, west of West End Park, May 21st. 1885.

Mr. F. A. Grout, of Columbia College, New York, who has just completed a monograph of *Brachythecium*, refers the above species to *B. rutabulum* (L.) var, *flavescens*, Brid.

147. **B. Donnellii,** Aust.

On limestone rocks in McKay's Woods, near the Lake, Sept. 25th, 1889.

148. **B. velutinum,** Bruch and Schimp.

On earth in Gilmour's Park, Chelsea, Que., Sept. 9th, 1889 ; on the bases of trees in Dow's Swamp, May 2nd, 1896.

149. **B. intricatum,** Hedw. (New to America.)

On rocks by a brook near Meeche's Lake, north of Chelsea, Que. Sept. 23rd, 1893.

150. **B. Starkii,** Bruch and Schimp.

On stones in the brook north of the Parry Sound Railway and west of West End Park ; on the bases of trees along the Beaver Meadow west of Hull, Que., Oct. 16th. 1884.

151. **B. œdipodium,** (Mitt.)

On old log's in woods Gilmour's Park, Chelsea, Que. Sept. 9th, 1889 ; on rocks in woods west of Hull, Que., Oct. 20th, 1891 ; on earth in Beechwood Cemetery, Oct. 12th, 1889.

152. **B. curtum,** Lindb.

On stones in damp places McKay's Woods, Oct. 12th, 1889.

153. **B. reflexum,** Bruch and Schimp.

On boulders along a brook near Meeche's Lake, north of Chelsea,
Que., Sept. 23rd, 1893.

154. **B. rivulare,** Bruch and Schimp.

On stones and roots in springs north of Beechwood Cemetery,
Oct. 12th, 1889.

155. **B. populeum,** Bruch and Schimp.

On rock's in Gilmour's Park, Chelsea, Que. Sept. 9th, 1889 ; on
rocks in McKay's Woods, Oct. 12th, 1889 ; and on boulders " Pine
Hill " Rockliffe Park, April 16th, 1896 ; on rocks at Meeche's Lake,
north of Chelsea, Que. Sept. 23rd, 1893.

156. **B. plumosum,** Bruch and Schimp.

On boulders in McKay's Woods, May 28th, 1884 ; on rocks in
woods, Meeche's Lake, north of Chelsea, Que. Sept. 23rd, 1893.

LIII. EURHYNCHIUM, Schimp.

157. **E. strigosum,** (Hoffm.) Bruch and Schimp.

On earth and stones in woods ; common. In McKay's Woods ;
on earth in Dow's Swamp, May 2nd, 1896 ; on old logs in Beechwood
Cemetery, Oct. 12th, 1889 ; on earth in woods at Meeche's Lake,Sept.
23rd, 1863.

158. **E. Novae-Angliæ,** (Lesq. and James.)

On old logs in woods at Meeche's Lake, Que. Sept. 23rd, 1893.

159. **E. Sullivantii,** (Spruce.) Lesq. and James.

On limestone rocks along the east side of the creek in the Beaver
Meadow west of Hull, Que. May 16th, 1885.

160. **E. hians,** (Hedw.) Lesq. and James.

On earth in woods near McKay's Lake, Oct. 12th, 1889. This
specimen is still doubtful.

LIV. RAPHIDOSTEGIUM, Lesq. and James.

161. **R. recurvans,** (Schwægr.) Lesq. and James.

Very common on the bases of leaning trees in woods. Beechwood Cemetery and McKay's Woods, April 28th, 1896 ; old woods at
Carleton Place.

LV. RHYNCHOSTEGIUM, Schimp.

162. **R. deplanatum,** Schimp.

On flat limestone rocks in McKay's Woods, May 2nd, 1885 ; also
on flat rocks and earth Carleton Place Sept. 26th, 1889 ; on earth in
Beechwood Cemetery, Oct. 12th, 1884.

163. R. pseudo-serrulatum, Kindb.

Leaves ovate or ovate-oblong, minutely denticulate, striate and chlorophyllose, not or indistinctly decurrent; cells lanceolate, the lower shorter and more dilated ; costa thin,vanishing above the middle. Stem leaves, short-acuminate or filiform-pointed ; branch-leaves with short subulate sometimes twisted point. Capsule arcuate; lid apicu-late ; pedicel rough very long. Stem subpinnate or irregularly divided. Tufts dark green, faintly glossy, radiculose at the base. Monœcious. Habit of *R. serrulatum.*

On earth and small stones in McKay's Woods near the lake, Oct. 12th, 1889,

164. R. serrulatum, (Hedw.)

On earth in McKay's Woods, Sept. 6th, 1889.

165. R. rusciforme, (Weis.) Schimp.

Abundent on stones in the brook that discharges Kingsmere four miles west of Chelsea, Que., Sept. 9th. 1889.

Var. inundatum, Kindb.

On rocks in the brook which discharges Kingsmere four miles west of Chelsea, Que., Sept. 9th, 1889 ; on rocks in a brook discharg-ing into Meeche's Lake, north of Chelsea, Que., Sept. 23rd. 1893.

## LVI. THAMNIUM, Schimp.

166. T. Alleghaniense (C. Muell.) Bruch and Schimp.

In crevices of thick bedded limestone rocks near McKay's Lake ; under ledges along the cliffs Rocklifle Park, April 22nd, 1896.

## LVII. PLAGIOTHECIUM, Schimp.

167. P. Passaicense, Aust.

On cedar stumps and trees in Dow's Swamp, Sept. 6th, 1889.

**168. P. turfaceum,** Lindb.

Common on stumps in Dow's Swamp,Oct. 20th, 1885 ; on stumps at Carleton Place, May 30th, 1884 : and on stumps " Pine Hill," Rockliffe Park, April 16th, 1896 ; on stumps west of Hull, Que., Oct. 24th, 1884.

**169. P. denticulatum,** (L.) Bruch and Schimp.

On the bases of trees at Britannia, Sept. 11th, 1890.; in the swamp north of Beechwood Cemetery, Oct. 12th, 1889 ; on stumps in the Swamp along Bank street, on the Glebe property, April 27th, 1896; on earth Eastman's Springs, Aug. 26th, 1890 ; on base of trees Carleton Place, Oct. 30th, 1884.

**170. P. sylraticum,** Bruch and Schimp.

On earth in damp woods borders of Dow's Swamp, Oct. 16th, 1884.; also in wet woods at Britannia, Sept. 14th, 1891.

**171. P. Sullivantiæ,** Schimp.

On earth on limestone ledges along the cliff north side of Rockliffe Park, May 7th, 1896 ; on earth at the roots of trees Meeche's Lake, north of Chelsea, Que., Sept. 23rd, 1893.

**172. P. membranosum,** Kindb.

Tufts dense, green and glossy. Leaves distichous, crowded and patent, flat. ovate-oblong, acute or short-acuminate, estriate, entire or denticulate above the middle, decurrent ; cells very long and narrow the alar large hyaline and subquadrate ; costa none or obsolete. Capsule cylindric-obovate, horizontally curved ; teeth yellow ; pedicel smooth, 2 cm. long. Lid unknown. Probably diœcious.

On dead wood in Dow's Swamp, Oct. 17th, 1884.

**173. P. brevipungens,** Kindb.

Tufts dense, dark green. Stems prostrate, irregularly pinnate ; branches attenuate. Leaves crowded, scarcely decurrent, ovate-oblong, acute or short-pointed, auricled, not plicate nor reflexed on the borders, entire or slightly denticulate at the apex ; upper cells long and narrow, the alar very distinct, quadrate, inflate and hyaline ; costa very short, thick and simple or none. Capsule curved ; lid short, conical ; pedicel smooth. Monœcious.

On stones in McKay's Woods opposite the entrance to Beechwood Cemetery, May 21st, 1885.

**174. P. attenuatirameum,** Kindb.

Tufts green, faintly shining, loose, with few rhizoids. Primary stem very short ; branches elongate, long-attenuate, finally flagelliform Leaves sub-distichous, the lower broadly ovate, obtuse or obtusate,

entire, long-decurrent, concave, recurved at the borders from the base
to the middle, at least at the one side ; cells chlorophyllose, somewhat
dilated, the lowest very much wider and shorter, nearly uniform ; costa
generally short and double, rarely simple and reaching to the middle ;
the other leaves gradually smaller, narrower and more acute or
acuminate.   Barren.

On rocks in Gilmour's Park, Chelsea, Que., Sept. 6th, 1889.

### LVIII. AMBLYSTEGIUM, Schimp.

**175. A. Sprucei,** Bruch. and Schimp.

On old cedar stumps in Dow's Swamp, Oct. 16th, 1884.

**176. A. subtile,** Bruch. and Schimp.

On the bases of large trunks in McKay's Woods, Oct. 10th, 1889 ;
on the borders of Dow's Swamp, Oct. 16th, 1884 ; in Beechwood
Cemetery, April 28th, 1896; also at Hog's Back, May, 1897.

**177. A. confervoides,** Bruch. and Schimp.

On flat limestone rocks in woods north of the Experimental Farm ;
along the Ottawa, west of Hull ; and in McKay's Woods near the
Lake, Sept. 9th, 1889 ; also on the cliffs and flat rocks, Rockcliffe Park,
April 16th, 1896.

**178. A. Yuratzskæ,** Schimp.

On dead wood, stones and bases of trees, McKay's Woods, Oct.
12th, 1889 ; on earth at Dow's Swamp, Oct. 16th, 1884 ; on stones in
Beechwood Cemetery, May 21st 1885.

**179. A. serpens,** Bruch. and Schimp.

On earth in woods and at the roots of trees ; common ; on roots of
trees, Rockcliffe Park, May 7th, 1896; at the base of trees, Hog's Back,
May, 1897.

**180. A. porphyhizum,** (Lindb.) Schimp.

On stones in McKay's Woods, Aug. 26th, 1889 ; also in woods at
Carleton Place, Sept. 25th, 1889.

**181. A. varium,** (Hedw.) Lindb.

On stones in damp parts of McKay's Woods and Beechwood
Cemetery, Oct. 12th, 1889 ; on stones at Leamy's Lake, Hull, Que.,
1896.

**182. A. fluviatile,** Bruch. and Schimp.

On stones in the brook that discharges Kingsmere, four miles west
of Chelsea, Que., Sept. 9th, 1889 ; on rocks in a brook at Meeche's
Lake, north of Chelsea, Que., Sept. 23rd, 1893.

183. **A. curvipes,** Guemb.

In pools in Dow's Swamp, Oct. 16th, 1884.

184. **A. adnatum,** (Hedw.) Lesq. and James.

On flat stones in McKay's Woods, May 16th, 1885 ; and in woods west of Hull, Que. ; on " Pine Hill," Rockcliffe Park, April, 16th, 1896.

185. **A. compactum,** (C. Muell.) Bruch. and Schimp.

On decayed wood in a swamp at Carleton Place, May 30th, 1884.

186. **A. riparium,** Bruch. and Schimp.

On sticks and roots in pools in woods near St. Patrick's Bridge ; also in the swamp north of Beechwood Cemetery, Oct. 12th, 1884.

## LIX HYPNUM, Linn.

187. **H. hispidulum,** Brid.

On the bases of trees around Ottawa ; on earth at Carleton Place, 30 miles from Ottawa, Sept.'25th, 1889 ; bases of trees, Gilmour Park, Chelsea, Que., Sept. 9th, 1889.

188. **H chysophyllum,** Brid.

On earth and roots of trees at Ottawa, Aug. 26th, 1889 ; common at Carleton Place, May 30th, 1884 ; on the cliff facing the Ottawa River, Rockcliffe Park, April 22nd, 1896.

189. **H. Sommerfeltii,** Myrin.

On earth at the bases of trees at Ottawa, Aug. 26th, 1889 ; on old logs near Leamy's Lake, Hull, Que., Nov. 9th, 1896.

190. **H. unicostatum,** C. M. and Kindb.

Differs from *Hypnum chrysophyllum* in the dense tufts, the stem more irregularly branching, creeping, the leaves shorter-acuminate, the alar cells smaller, not yellow, the costa more distinct, the capsule smaller, the perichetial leaves gradually acuminate-subulate or filiform-pointed with acumen arcuate.

On earth in the Mer Bleue, close to the sulphur spring, near Eastman's Springs, Oct. 4th, 1890.

191. **H. Kneiffii,** (Bruch. and Schimp.)

Bogs and swamps. In swamps north of Beechwood Cemetery, Oct. 22th, 1889 ; also in Dow's Swamp ; in a bog at Britannia, six miles west of Ottawa, Sept. 14th, 1891; in a swamp near Hog's Back, 1897.

Var. **laxum,** Kindb.

In Dow's Swamp, Oct. 12th, 1884.

Var. **rectifolium**, Kindb.

On damp rocks in McKay's Woods, Aug. 26th, 1889.

Var. **platyphyllum**, Kindb.

Leaves very broad and short-acuminate.

On damp rocks in woods east of the end of the Electric Road, Rockcliffe Park, Aug. 26th, 1889.

192. **H. Sendtneri**, Schimp.

In water in Dow's Swamp ; in a bog at Britannia, six miles west of Ottawa, Sept. 14th, 1891.

193. **H. fluitans**, Linn.

In a swamp along the Beaver Meadow west of Hull, Que., Oct. 16th, 1891 ; in boggy places north of Beechwood Cemetery and McKay's Lake, Sept. 25th, 1889.

194. **H. uncinatum**, C. M. and Kindb.

Quite common in wet woods north of Beechwood Cemetery, Oct. 12th, 1889; in Dow's Swamp and at the Hog's Back, May 22nd, 1897; also in the Mer Bleue, Oct. 4th, 1890.

195. **H. conflatum**, C. M. and Kindb.

Allied to *Hypnum Kneiffii*. Stem slender, subfiliform, distantly pinnate, not radiculose. Leaves small, concave, distant, denticulate all around ; stem leaves decurrent, from a broad-ovate base suddenly narrowed into a very short, subulate-filiform, straight point ; alar cells very large, hyaline or faintly yellowish, the other nearly uniform, oblong-lanceolate, conflate ; costa pale-yellow, vanishing in the acumen ; branch leaves narrower, oblong-lanceolate, more or less short-acuminate, curved or straight. Capsule very small, arcuate and contracted below the mouth. Diœcious.

In a swamp near Britannia, six miles west of Ottawa, Sept. 14th, 1891.

196. **H. rugosum**, Linn.

On dry rocks along the exposed cliffs facing the Ottawa River, Rockcliffe Park, April 22nd, 1896.

197. **H. crista-castrensis**, Linn.

On logs and on earth in damp woods around Ottawa. Collected near McKay's Lake, April 28th, 1896 ; in Dow's Swamp and at Casselman.

198. **H. molluscum**, Hedw.

On old logs in Dow's Swamp, Sept. 9th, 1889 ; on logs near the Lake in McKay's Woods, April 28th, 1896.

199. **H. reptile**, Michx.

Abundant on trunks of trees in woods around Ottawa. On trunks "Pine Hill," Rockcliffe Park, April 14th, 1896 ; on old logs in Beechwood Cemetery and along the Aylmer Road, Oct. 12th, 1887.

200. **H. pallescens**, Schimp.

On limestone rocks on an old stone fence along the Aylmer Road, near Tetreauville, west of Hull, Que., Sept. 24th, 1891 ; on boulders along the road leading east from the end of the Electric Road, Rockcliffe Park, May 7th, 1896.

201. **H. Canadense**, Kindb.

On old logs and stones at Rockcliffe near the end of the Electric Road, Rockcliffe Park, May 7th, 1886 ; on stones in woods west of Hull, Que., Sept. 11th, 1891 ; on rocks in Rockcliffe Park, Sept. 25th, 1889.

202. **H. fertile**, Sendt.

On old logs at Casselman, June 11th, 1892.

203. **H. imponens**, Hedw.

Common on rotten logs around Ottawa, at Chelsea and Casselman.

204. **H. arcuatiforme**, Kindb.

Tufts dense, green, not glossy. Stem creeping, subpinnate. Leaves arcuate, ovate-lanceolate, generally short-acuminate or sub-obtusate, entire, decurrent, not striate ; alar cells large, well-defined, orange, the other pale and narrow ; costa none or short and double. Capsule subcylindric, curved, not striate nor furrowed, constricted below the wide mouth ; teeth when dry incurved, pale-yellow, hyaline-margined ; cilia long, appendiculate ; pedicel about 3 cm. long. Probably diœcious. Resembles in habit *Hypnum cupressiforme.* Lid and male flowers not found.

The allied *Hypnum Lindbergii*, Mitt. (*H. arcuatum*, Lindb.) differs at once in the not creeping, irregularly divided stem, the shorter pedicel, the larger capsule, &c.

On earth near the gate of Beechwood Cemetery, Sept. 29th, 1889

205. **H. Renauldii**, Kindb.

Agrees with *Hypnum curvifolium* in the stem being more or less pinnate, the inner basal leaf-cells finally vellow ; with *Hypnum Lindbergii* in the leaves being decurrent, alar cells very much dilated, he capsule not plicate in a dry state ; differs from both in the entire leaves. *Hypnum pratense* differs in the leaves not being striate nor decurrent, and alar cells not evolute.

On earth near the small lake at the head of the Beaver Meadow, west of Hull. Que., May 14th, 1891.

### 206. H. curvifolium, Hedw.

On old logs in damp or swampy woods around Ottawa.

### 207. H. Lindbergii,' Mitt.

On earth on rocks along Leamy's Lake, Hull, Que., Nov. 9th, 1896.

### 208. H. Haldanianum, Grev.

Abundant on rotten logs around Ottawa. In woods at Leamy's, Lake, Chelsea, Ironsides and Meeche's Lake, Que. ; in woods at the south end of Cowley's Farm, west of Hintonburgh, April 18th, 1896 ; on logs in Beechwood Cemetery, Oct. 12th, 1889 ; rotten logs at Casselman, Aug. 20th. 1884.

### 209. H. pseudo-drepanium, C. M. and Kindb.

Tufts loose, green, faintly glossy. Secondary stems very long, flaccid, subpinnate, sparingly radiculose, faintly compressed ; paraphyllia none ; branchlets very short, curved at the apex. Leaves plicate, entire, from a short broad-ovate base narrowed into a short-incurved, acute acumen, crowded, loosely appressed when dry, not decurrent, not distinctly chlorophyllose ; basal cells hyaline, dilated (the inner sometimes yellowish (thick-walled, the alar large sub-rectangular, well-defined, the other longer and narrower ; auricles excavate ; costa indisdinct or short and double. Dioecious. Female plants not found.

On old logs in Dow's Swamp, Sept, 16th, 1889.

### 210. H. stramineum, Dicks.

In the Mer Bleue, near Eastman's Springs.

### 211. H. cordifolium, Hedw.

In wet woods and swamps ; in holes in Dow's Swamp ; in pools in the swamp north of Beechwood Cemetery, May 12th, 1897 ; swamp west of Hull, Oct. 16th, 1884.

### 212. H. Richardsoni, (Mitt.) Lesq. and James.

In water along the south end of McKay's Lake, near Beechwood Cemetery, Sept. 26th, 1893.

### 213. H. giganteum, Schimp.

In the spring creek in Dow's Swamp ; also in the Beaver Meadow creek, west of Hull, Que., Oct. 3rd, 1893.

**214. H. Schreberi,** Willd.

Abundant on earth in old woods around Ottawa. In shady woods, Rockcliffe Park, April 22nd, 1896.

### LX. HYLOCOMIUM, Schimp.

**215. H. splendens,** (Hedw.) Schimp.

Very abundant in damp woods everywhere around Ottawa. In shady woods along the cliff facing the Ottawa River, Rockcliffe Park, April 22nd, 1896 ; Casselman, Aug. 20th, 1884.

**216. H. brevirostre,** Schimp.

On damp rocks near High Falls, Lievre River, near Buckingham, Que., May 19th, 1884.

**217. H. triquetrum,** (Linn ) Schimp.

Very common in old cool or damp woods. In shady woods along the cliff facing Ottawa River, Rockcliffe Park, April 22nd, 1896 ; Casselman, Aug. 20th, 1884.

**218. H. pyreniacum,** (Spruce) Lind.

On rocks by a brook, Meeche's Lake, north of Chelsea, Que., Sept. 23rd, 1893      (*Addendum.*)

**219. Didymodon trachneuron,** Kindb. (Just published)

Leaves short-attenuate, subacute, brittle ; costa pale, papillose at back, sometimes excurrent. Tufts low, green. Capsules unknown.

On large boulders in a brook near North Wakefield, Que., Sept. 13th, 1893. (J. M. Macoun)

**220. Grimmia arctophila,** Kindb.

Leaves ovate oblong, involute above, suberect when moist, appressed when dry, the uppermost not larger, basal cells generally short, the upper short-opaque ; hair point not long; capsule sub-oblong, brownish ; lid apiculate ; annulus indistinct ; pedicel straight. Tufts generally cohering, blackish when dry, about 3 cm. high.

On rocks, Paugan Falls, on the Gatineau River, Que , Aug. 24th, 1894.

### HEPATICÆ (LIVERWORTS).

#### I. FRULLANIA, Raddi.

**221. F. Eboracensis,** (Gottsch.) Lehm.

Very common, especially on beech trees in all the old woods around Ottawa. On bark of various trees, especially beech at Casselman, Eastman's Springs. Carleton Place, Chelsea and near Hull ; on beech trunks, Pine Hill, Rockcliffe Park, April 16th, 1896 ; Beechwood Cemetery on beech trees, Oct. 14th, 1884 ; on trees at Leamy's Lake, Hull, Que., Sept. 6th, 1889.

## II. LEJEUNEA, Libert.

**222. L. calcarea,** Libert.

On the bases of cedar (*Thuya occidentalis*) trees in Dow's Swamp, Sept. 21st, 1889 ; on cedar bark in woods, Rockcliffe Park, April 22nd, 1896. Very rare.

## III. RADULA, Dumortier.

**223. R. complanata,** Dumort.

Very common on rocks and the roots and bases of trees in woods around Ottawa. On rocks and trees in rear of Cowley's Farm, west of Hintonburg, at Chelsea, Kingsmere, Meeche's Lake, and near Hull, Que on trees at Britannia and Carleton Place ; very common in Beechwood Cemetery and McKay's Woods, April 14th, 1896 ; on rocks along the Ottawa River, Rockcliffe Park, Oct. 16th, 1889.

## IV. PORELLA, Dill.

**224. P. platyphylla,** (L.) Lindb.

Very common on the bases of trees by brooks and on ledges of damp limestone rocks around Ottawa. On trees by the brook in rear of Cowley's Farm,west of Hintonburg,April 18th, 1896 ; also on trees at Leamy's Lake, Hull, and on rocks near Tetreuville, west of Hull, Que., Oct. 12th, 1887 ; on trees in the swamp, Beechwood Cemetery, Aug. 24th, 1884.

## V. PTILIDIUM, Nees.

**225. P. ciliare,** Nees.

Very common on old logs and rocks. On old logs, Beechwood Cemetery, McKay's Woods, and Rockcliffe Park, April 14th, 1896 ; in woods at Carleton Place, Eastman's Springs and Casselman ; Chelsea and Kingsmere, also Leamy's Lake, Hull, Que., Oct. 12th, 1884.

## VI. TRICHOCOLEA, Dumortier.

**226. T. tomentella,** (Ehrh.) Dumort.

In moss in a swamp at Meeche's Lake, north of Chelsea, Que., Sept. 23rd, 1893 ; in moss in Dow's Swamp, May 2nd, 1896.

## VII. BAZZANIA, S. F. Gray.

**227. B. trilobata,** S. F. Gray.

On old logs and stumps in swamps and wet woods around Ottawa.

In Dow's Swamp, Oct. 12th, 1884 ; and on old logs, McKay's Woods, April 28th, 1895.

## VIII. LEPIDOZIA, Dumortier.

### 228. L. reptans, Dumort.

On old logs and stumps in cedar and other swamps. In the swamp near the head of the Beaver Meadow, west of Hull, Que., April 18th, 1896 ; in Dow's Swamy, May 2nd, 1896 ; on old logs, Cassselman, Aug. 20th, 1884.

## IX. BLEPHAROSTOMA, Dumortier.

### 229. B. trichophyllum, Dumort.

On very rotten logs and stumps in swamps and wet woods ; also earth and stones. On rotten wood in Dow's Swamp, May 2nd, 1896 ; on olc logs in Beechwood Cemeterá, May 8th, 1895 ; in a swamp at Casselman, June 18th, 1894.

## X. CEPHALOZIA, Dumortier.

### 230. C. Virginiana, Spruce.

On old logs in damp woods. On rotten wood in Beechwood Cemetery, May 9th, 1885.

### 231. C. multiflora, Spruce.

Very common on rotten wood in damp woods and swamps. On rotten wood in Dow's Swamp, May 2nd, 1896 ; also in the swamp on the Glebe property, Bank St., Ottawa, April 27th, 1896 ; on stones and old logs near Experimental Farm, April 16th, 1892 ; on old logs, Beechwood Cemetery, May 9th, 1885 ; on dead wood, Moose Creek, Sept. 5th, 1891 ; also at Eastman's Springs, Sept. 29th, 1892.

### 232. C. divaricata, Dumort.

On dead logs in a swamp north of the Experimental Farm, April 16th, 1892.

### 233. C. pleniceps, (Aust.)

On rotten wood and on stones around Ottawa. Rotten wood in the swamp opposite the entrance to Beechwood Cemetery, Oct. 12th, 1884 ; on stones in the brook by McKay's Lake, Aug. 26th, 1884.

### 234. C. bicuspidata, Dumort.

On rotten wood in Dow's Swamp, May 2nd, 1896 ; rotten wood in the swamp opposite the entrance to to Beechwood Cemetery, May 9th, 1885.

235. **C. curvifolia,** (Dicks.) Dumort.

On old logs in the swamp west of the old entrance to Beechwood Cemetery, May 8th, 1885.

### XI. ODONTOSCHISMA, Dumortier. ·

236. **O. denudata,** (Mart.) Lindb.

On a pine log in Beechwood Cemetery, May 9th, 1895.

### XII. KANTIA, S. F. Gray.

237. **K. Trichomanis,** (L.) S. F. Gray.

On dead wood in swamps and wet woods around Ottawa. Damp woods, Carleton Place, 30 mmiles west of Ottawa, May 20th, 1884 ; on dead wood in a swamp north of the Experimental Farm, April 16th, 1892.

### XIII. SCAPANIA, Dumortier.

238. **S. glaucocephala,** (Tayl.) Aust.

On elm and other logs at Ottawa. Along the Beaver Meadow, Hull, Que., Sept. 23rd, 1883 ; on old logs, Dow's Swamp, Oct. 20th, 1884.

239. **S. curta,** (Mart.)

On logs in Rockcliffe Park, Sept. 24th, 1897 ; in Dow's Swamp on old logs, Oct. 5th, 1884.

240. **S. nemorosa,** (Linn.) Dumort.

On rotten logs by a brook at Meeche's Lake, north of Chelsea, Que., Sept. 23rd, 1893 ; on earth on a wet bank by the Rideau River at the Hog's Back, May 14th, 1897.

### XIV. GEOCALYX, Nees.

241. **G. graveolens,** (Schrad ) Nees.

On dead wood and old stumps in the swamp on the Glebe property, Bank St., Ottawa, April 27th, 1896 ; on earth in a swamp north of the Experimental Farm, April 16th, 1892 ; on rotten logs in Dow's Swamp, Oct. 16th, 1885 ; on dead logs near Leamy's Lake, Hull, Que., April 23rd, 1892.

### XV. LOPHOCOLEA, Dumortier.

242. **L. bidentata,** (Linn.) Dumort.

On rotten logs at Eastman's Springs, 12 miles from Ottawa, Sept. 27th, 1892.

**243. L. minor, Nees.**

On limestone rocks and calcareous earth, Rockcliffe Park, April 22nd, 1896 ; on rocks at Carleton Place, May 30th, 1884 ; on dead wood in Dow's Swamp, Sept. 4th, 1889 ; on rocks north of Experimental Farm, April 16th, 1892.

**244. L. Macounii, Aust.**

On rotten wood, in woods west of the Beaver Meadow, Hull, Que., Sept. 23rd, 1883.

**245. L. heterophylla, (Schrad.)**

Quite common on rotten wood in swamps and wet woods at Ottawa. On rotten wood at Hintonburg, April 18th, 1896 ; on elm logs, Moose Creek, Sept. 6th, 1891 ; on rocks north of Experimental Farm, April 16th, 1892 ; on old logs, Beechwood Cemetery, April 23rd, 1892 ; by Leamy's Lake, Hull, Que., Sept. 29th, 1889.

### XVI. PLAGIOCHILA, Dumortier.

**246. P. porelloides, (Torrey.) Lindb.**

Common on earth and old logs along brooks at Ottawa. On old logs in the swamp near the entrance to Beechwood Cemetery, April 28th, 1896 ; on logs in Dow's Swamp, 16th Oct., 1889 ; on damp rocks, Meeche's Lake, Hull, Que., Sept- 24th, 1893.

**247. P. asplenioides, (Linn.) Dumort.**

On earth in wet woods at Ottawa, April 24th, 1885 ; on the margin of the brook west of Victoria Park, May 9th, 1884 ; on wet rocks in a brook at Meeche's Lake. Que.. Sept. 23rd, 1893.

**248. P. interruptus, (Nees.) Dumort.**

On earth in Dow's Swamp, Sept. 24th, 1889.

### XVII. HARPANTHUS, Nees.

**249. H. scutatus, (Web. and Mohr.) Nees.**

On rotten logs in McKay's Woods, April 28th, 1896 ; also in woods at Leamy's Lake, Hull, Que , 16th Oct., 1892.

### XVIII. JUNGERMANNIA, Micheli.

**250. J. Schraderi, Mart.**

On dead logs in damp woods and swamps around Ottawa. On old logs in the swamp in rear of Cowley's Farm, west of Hintonburg, April 18th, 1896 ; Dow's Swamp, 1885 ; at Carleton Place, 1884 ; on old logs, Experimental Farm. April 16th, 1891.

251. **J. barbata,** Schreb.

On damp limestone cliffs facing the Ottawa River, Rockcliffe Park, May 7th, 1896 ; on rocks, Gilmour's Park, Chelsea, Que, Sept. 9th, 1889 ; on rocks near Ironsides and on rocks Meeche's Lake, Que., Sept. 23rd, 1893.

252. **J. attenuata,** Lindenb.

On rocks on damp cliffs, Rockcliffe Park, April 16th, 1891.

253. **J. lycopodioides,** Wallr.

On rocks on the east side of the cliff, close to the old sawmill, Rockcliffe Park, Oct. 26th, 1889.

254. **J. exsecta,** Schmid.

Common on dead wood, Beechwood Cemetery, April 23rd, 1892 ; on moss on logs in Dow's Swamp, Oct., 1884.

255. **J. incisa,** Schrad.

On rotten wood in Beechwood Cemetery, Sept. 2nd, 1884 ; also on old logs west of Beaver Meadow, Hull, Que., Oct. 6th, 1885.

256. **J. excisa,** Dicks.

Rather common on rotten wood around Ottawa ; in Dow's Swamp, Oct., 1884 ; also at Meeche's Lake near Chelsea, Que., Sept. 24th, 1893.

257. **J. pumila,** With.

On rocks along Meeche's Lake, north of Chelsea, Que., Sept. 23rd, 1893.

### XIX. FOSSOMBRONIA, Raddi.

258. **F. Dumortieri,** Lindb.

On earth subject to inundation close to Leamy's Lake, Hull, Que., Sept. 24th, 1889.

## XX. BLASIA, Micheli,

**259. B. pusilla,** (Linn.)

On earth subject to inundation in a gully at Leamy's Lake, Hull, Que., Sept. 24th, 1889 ; on wet clay banks, Meeche's Lake, north of Chelsea, Que., Sept. 23rd 1893 ; also on clay banks at the confluence of the Leivre River with the Ottawa River near Buckingham, Que., Sept. 26th, 1892.

## XXI. PELLIA, Raddi.

**260. P. epiphylla,** Corda.

On earth subject to inundation in a gully at Leamy's Lake, Hull, Que., Sept. 24th, 1889.

## XXII. ANEURA. Dumortier.

**261. A. latifrons,** (Lindb.) Dumort.

On old logs in the swamp, Glebe property, Bank St., Ottawa, April 27th, 1896 ; on old logs, Beechwood Cemetery, April 23rd, 1892.

**262. A. palmata** (Hedw.)

On old logs and stumps in Dow's Swamp, Oct. 6th. 1885 ; also on logs in McKay's Woods, Oct. 17th, 1890.

**263. A. sessilis,** (Sprengel.) Dumort.

On old logs in a swamp about a mile south-east of Carleton Place ; in fine fruit, May 30th, 1884.

**264 A. pinguis,** (Linn.) Dumort.

Amongst peat moss in the Mer Bleue, near Eastman's Springs, June 15th, 1892.

## XXIII. METZGERIA, Raddi.

**265. M. myriopoda,** Lindb.

On damp rocks near Ironsides, five miles north of Hull, Que., Oct. 21st, 1884·

**266 M. conjugata,** Lindb.

On stones in McKay's Woods near the Lake, Oct. 9th, 1884.

## XXXIV. ANTHOCEROS, Micheli.

**267. A. Macounii,** Howe (N. Sp.) Torr. Bull.    Vol. xxv, page 19 (1898).

Thallus forming small dark green rosettes, 4-10 mm., in diameter, strongly undulate-crisped, subradiately inciso-laciniate or somewhat broadly lobed, rugose, pitted, sometimes slightly lamellate ecostate 6-8 cells thick in axile parts, cavernose, becoming at the margin gradually 3 or 2-stratose, now and then glandular-thickened ; surface cells distinct, translucent, lightly protuberant, subrhombic, trapezoidal, or oblong-pentagonal 35·75 x 30·35 mik. ; *Nostoc* colonies spherical ; monoicous ; antheridia in groups of 3 or 4 ; involucres short, sometimes united in pairs cylindrical oblong, or by contraction at base and mouth dolioform or subglobose, ·85-1·25 x ·5-·9 mm., incrassate except at the thin erose or subentire mouth ; capsule black, erect or a little curved, 3·6 x 3·5 mm., thick-walled, with numerous stomata, the valves rigid or slightly flexuose when dry, brittle and often broken ; columella sometimes appendiculate ; spores fuscus or black, rounded-tetrahedral densely and rather minutely muriculate on both the inner and outer faces, 48-65 mik. in maximum diameter ; sterile cells short, nearly as broad as long, without spiral thickenings, separate or variously adherent, often shriveled and inconspicuous.

On earth subject to inundation along the discharge of Leamy's Lake, near Hull, Que., Sept. 24th, 1891.

### XXV. MARCHANTIA, Marchant.

**268. M. polymorhpa,** Linn.

Quite common around springs and on earth along the borders of swamps around Ottawa.   On earth, by the lake in McKay's Woods, April 28th, 1896.

### XXVI. PREISSIA, Nees.

**269. P. commutata,** Nees.

Under dripping limestone rocks under the cliffs near the old mill, east side of Rockcliffe, May 7th, 1896.

### XXVII. CONOCEPHALUS, Necker.

**270. C. conicus,** Neck.

Quite common on old logs and earth by brooks around Ottawa. In a swamp on the rear of Cowley's Farm, west of Hintonburg, April 18th, 1896.

### XXVIII. GRIMALDIA, Raddi.

**271. G. rupestris,** (Nees.) Lindenb.

On calcareous earth in crevices of rocks near Governor's Bay, Rockcliffe Park, May 20th, 1884.

### XXXIX. RICCIA, Micheli.

**272. R. arvensis,** Aust.

On damp earth covered by the spring floods around the east side of Leamy's Lake, Hull. Que., Sept. 24th, 1889.

**273. R. fluitans,** Linn,

Very abundant in Patterson's Creek, Bank St., Ottawa ; also in the Beaver Meadow Creek, west of Hull, Que., Oct. 9th, 1896.

**274. R. natans,** (Linn.) Corda.

In stagnant pools east of Beechwood Cemetery, April 23rd, 1892.

**Forma terrestris.**

Grows late in the season where pools had been in the spring. On earth along the Ottawa and Lievre rivers, near Buckingham, 20 miles below Ottawa, Sept. 18th, 1892 ; also along the discharge of Leamy's Lake, Hull, Que., Oct. 16th, 1893.

## LICHENES.

### I. RAMALINA, Ach.

**275. Ramalina calicaris** (L) var. **fastigiata,** Fr.

On old rails, cld logs and trunks rather rare. On bark of young red maples at Britannia, April 20th, 1895 ; rare on old logs near Ottawa River, Hull : on a pine tree, Pine Hill, Rockcliffe Park, rare on old stumps and rails in a fence one mile south east of Billing's Bridge ; also on red maples Leamy's Lake ; on trees in a swamp in Stittsville ; old logs King's Mountain, west of Chelsea.

Var. **farinacea,** Schacr.

On trees, old fence rails, and ledges of rocks. Rare on bark along the C.A.R. in Stewart's Bush, April 12th, 1895 : on old fence rails West End Park ; on limestone ledges on the face of the cliff, opposite Gatineau Point, Rockcliffe Park.

Var. **canaliculata,** Fr.

On a balsam fir in a swamp a little east of Stittsville, May 14th, 1897.

**276. Ramalina pusilla** (Prev.) Var. **geniculata,** Tuck.

On twigs of spruce tress near Ironsides, Que., Oct. 6th, 1891.

### II. CETRARIA, (Ach.) Fr.

**277. Cetraria ciliaris,** (Ach.)

On old fence rails and boards and occasionally on pine stumps and trees. On old fence rails, West End Park, April 16th, 1892 ; on old board fencing at Buckingham, Que. ; on an old pine stump one mile above Britannia ; on tamarack trees in a swamp at Stittsville, May 14th, 1897.

278. **Cetraria sæpincola**, (Ehrh.) Ach.

Rare. Occasionally in swamps. On branches of black spruce in the Mer Bleue, Eastman's Springs, June 16th, 1891.

279. **Cetraria lacunosa**, Ach.

Rare in the Ottawa district. ; on trees and rails. On old fence rails and boards at Buckingham, Que., May 14th, 1896.

280 **Cetrartia Oakesiana**, Tuckerm.

Very rare in Canada. On the base of living pine trees and at the base of pine stumps, Rockcliffe Park, April 17th, 1895 ; on the base of a pine stump by a swamp at Stittsville : old pine log, King's Mountain, west of Chelsea, May 22nd, 1897,

281. **Cetraria juniperina** (L.)   Var. **Pinastri**, Ach.

Rare in the Ottawa district. On dead braiches of black spruce and old logs in the Mer Bleue, at Eastman's Springs, June 16th, 1891 ; on branches of tamarack in a swamp at Stittsville, May 14th, 1897.

### III. EVERNIA, Ach.

282. **Evernia prunastri**, (L.) Ach.

On trees, stumps and old fences ; rare. On old pine stumps at Britannia, April 10th, 1884 ; on old rails along the Richmond Road above Hintonburg ; on an old fence, Ottawa East ; on trees in the swamp west of Hull Station ; on branches of tamarack trees in a swamp at Stittsville, May 14th, 1897.

### IV. USNEA, (Dill.) Ach.

283. **Usnea barbata**, (L.) Var. **hirta**, Fr.

On trunks south of the Aylmer Road, west of Hull, Que., April 26th, 1891 ; on a spruce tree, Rockcliffe Park ; on pine stumps at Britannia ; on spruce and tamarack trees in the Mer Bleue, at Eastman's Springs ; on tamarack trees in a swamp at Stittsville ; old log, King's Mountain, west of Chelsea, May 22nd, 1897.

### V. ALECTORIA, (Ach.) Nyl.

284. **Alectoria jubata**, (L.)   Var. **chalybeiformis**, Ach.

Rare on dead wood or on the earth. On old pine stumps at Britannia, April 20th, 1895 ; on tamarack trees in a swamp at Stittsville, May 14th, 1897.

Var. **implexa**, Fr.

Quite common in tamarack and other swamps, hanging like black hair from the branches. On black spruce and tamarack in the Mer Bleue, at Eastman's Springs, June 16th, 1891.

## VI. THELOSCHISTES, Norm.

**285. Theloschistes polycarpus, (Ehrh.)**

A common species on living trees and dead wood. On black ash and white cedar trunks and balsam poplar branches in Stewart's Bush, April 12th, 1895 ; common on willow, red ash and alder at Britannia ; on balsam poplar and white ash trunks, at Hintonburg ; and on ash and rock elm in Beechwood ; on old fence boards at Billing's Bridge ; on red maple and ash at Leamy's Lake ; rare on trees at Stittsville ; on trunks, King's Mountain, west of Chelsea, May 22nd, 1897.

**286. Theloschistes concolor, Dicks.**

On ash trees west of West End Park, April 16th, 1892 ; on white cedar bark by the C.A.R. in Stewart's Bush ; quite common on willow, ash, maple and alder at Britannia ; common on bark of trees, Aylmer Road, west of Hull ; on trunks of black ash in woods west of Hintonburg ; also on basswood trees at Carleton Place ; on a black ash log, Ottawa East, and on ash trees in Beechwood : on old fence boards at Billing's Bridge ; on various trees in woods, Leamy's Lake ; on trees at Stittsville ; on trunks, King's Mountain, west of Chelsea, May 22nd. 1897.

## VII. PARMELIA, (Ach) De Not.

**287. Parmelia perlata, (L.) Ach.**

Not uncommon on trunks in wet woods or swamps. A fine species but seldom found in fruit. On ash trees by the C A.R., Stewart's Bush, April 12th, 1895 ; on a spruce trunk in Rockcliffe Park ; on birch trees, Skead's Far n, Richmond Road ; on white cedar, black ash, and cherry birch in the swamp near Beechwood Cemetery ; on a birch tree in Dow's Swamp ; on trees in woods at Chelsea, Que. ; also on trees at Carleton Place ; on trees in the swamp west of Hull Station, Oct. 9th, 1896.

**288. Parmelia tiliacea, (Hoffm.) Floerk.**

Rather uncommon except in deep, cool woods. On birch trees on Skead's Farm, Richmond Road ; on beech and other trees, Rockcliffe Park ; on young spruce trees in woods, Beaver Meadow west of Hull ; on a beech tree in woods one mile south-east of Billing's Bridge ; on red maple at Leamy's Lake ; on beech trunks, King's Mountain, west of Chelsea, May 22nd, 1897.

**289. Parmelia Borreri, (Turn.)**

Apparently rare in the vicinity of Ottawa. On trunks in woods north of Beechwood Cemetery, April 23rd, 1891 ; on trees in a swamp at Stittsville, May 14th, 1897.

Var. **rudecta**, Tuckerm.

Very common on old rails and dead wood, around Ottawa.  On dead wood and old rails in Stewart's Bush, April 12th, 1895 ; on dead wood, living spruce trees and boulders, Rockcliffe Park ; on dead wood at Britannia ; on dead pines, Aylmer Road, west of Hull ; on white cedar north of Beechwood ; on dead trees at Carleton Place ; on a beech tree in woods one mile south-east of Billing's Bridge ; on large trees in woods, Leamy's Lake ; on old rails and logs, King's Mountain, west of Chelsea, May 22nd, 1897.

**290. Parmelia saxatilis,** (L.) Fr.

On trunks, dead wood, and rocks.  On trunks in woods at Leamy's Lake, May 7th, 1897 ; on tamarack and other trees in a swamp at Stittsville, May 22nd, 1897.

Var. **sulcata**, Nyl.

On trunks, dead wood and rocks.  On the branches of a dead spruce, Rockcliffe Park, April 17th, 1895 ; on boulders along a fence, Ottawa East ; on red maples in woods near Leamy's Lake, Hull, Que. ; on old logs and rails east of Stittsville ; on rocks and trunks, King's Mountain, west of Chelsea, May 22nd, 1897.

**291. Parmelia physodes,** (L.) Ach.

On dead wood, old fence rails, boards and rocks.  On the branches of dead spruce, Rockcliffe Park, April 17th, 1895 ; on pine stumps at Britannia ; on old fence rails and boards at Buckingham, Que. ; on old fence rails and tamarack trees at Stittsville ; on old logs and rails, King's Mountain, west of Chelsea.

**292. Parmelia colpodes,** (Ach.) Nyl.

Not rare, chiefly on tamarack trees.  In a swamp a little east of Stittsville, north of the Can. Pac. Railway, May 14th, 1897 ; on trunks, western slope King's Mountain, May 22nd, 1897.

**293. Parmelia olivacea,** (L.) Ach.

On trees and old rails.  On alders, red maple and red ash at Britannia, April 20th, 1895 ; rare on pine trees west of Hull ; on old rails, Ottawa East ; on old rails at Dow's Swamp ; and on young pines at Carleton Place ; on old pine stumps, Leamy's Lake ; common on tamarack trees in a swamp at Stittsville, May 14th, 1897.

Var. **aspidiota**, Ach.

Same habitat as the species.  On alder bushes at Britannia, April 20th, 1895 ; on alder bushes near Leamy's Lake, Hull, Que. ; on tamarack trees in a swamp at Stittsville, May 14th, 1897.

Var. **sorediata**, (Ach.) Nyl.

On trees and rocks ; rare. On maple trunks north of Aylmer Road, west of Hull, Que., April 26th, 1891.

294. **Parmelia caperata**, (L.) Ach.

On trunks, dead wood, and stones ; common. On old rails and pine trees, Clemow's Woods, Bank St., April 12th, 1895 ; common on dead and living trees at Rockcliffe, Beechwood and Ottawa East ; abundant on old rails and dead wood at Britannia ; common on dead pines Aylmer Road and by the Beaver Meadow, Hull ; old fence rails, west of Hintonburg and West End Park ; on rails in Dow's Swamp ; and on trees at Carleton Place ; on old fences around Billing's Bridge ; on trees of all kinds at Leamy's Lake ; on old stumps and fences at Stittsville ; very common, King's Mountain, Chelsea, Que.

295. **Parmelia conspersa**, (Ehrh.) Ach.

Abundant on boulders in all old fields and fences around Ottawa. Collected in Rockcliffe Park, Ottawa East, by Dow's Swamp, fields at Hintonburg, and along the Aylmer Road west of Hull ; on boulders around Billing's Bridge ; on boulders at Brigham's Creek, near Leamy's Lake ; on boulders at Stittsville ; on boulders and other rocks, King's Mountain, Chelsea, Que.

## VIII. PHYSICA, DC.

296. **Physcia speciosa**, (Ach.) Nyl.

On trees and mossy rocks in woods. On trees at Ottawa, 1884 ; on trees in Beechwood Cemetery ; on trunks, Pine Hill, Rockcliffe Park, April 16th, 1896 ; on a hemlock trunk in woods north of Beechwood Cemetery ; on a beech tree in woods one mile south-east of Billing's Bridge ; on the bases of basswood trees in woods at Leamy's Lake, May 7th, 1897.

297. **Physica granulifera**, (Ach.) Tuckerm.

On trunks. On bark of trees north of Aylmer Road, Hull, Que., April 26th, 1891 ; on ash trees, Cowley's Farm, west of Hintonburg ; on large trees in woods at Leamy's Lake, May 7th, 1897.

298. **Physcia pulverulenta**, (Schreb.) Nyl.

On trunks and rocks. On black ash trunks, Stewart's Bush, April 12th, 1895 : on ash trees at Britannia ; on living and dead trees, Skead's Farm, Hintonburg ; quite common on ash and other trees along the Aylmer Road west of Hull ; on ash trunks in Dow's Swamp, Ottawa East, Beechwood and Rockcliffe Park ; on large trunks in woods, Leamy's Lake : on trees at Stittsville ; on trunks and rails, King's Mountain, Chelsea, Que., May 22nd, 1897.

### 299.  Physcia stellaris, (L.)

Very common on trunks and dead or living branches.  On black ash trees in a swamp in Stewart's Bush, April 12th, 1895 ; on dead and living trees at Rockcliffe Park, Beechwood, Ottawa East, Dow's Swamp and Skead's Farm, Richmond Road, and common at Britannia ; common on trees in woods at Leamy's Lake ; on balsam trees at Stittsville ; on trunks and rails, King's Mountain, Chelsea, Que., May 22nd, 1897.

Var. aipolia, Nyl.

Same range as the species.  On the bark of trees at Ottawa, 1884 ; on a black ash trunk in a swamp west of Beechwood Cemetery ; on beech and maple trees in woods one mile south-east of Billing's Bridge ; on trees in woods, Leamy's Lake ; on tamarack and other trees at Stittsville, May 14th, 1897.

### 300.  Physcia astroidea, (Fr.) Nyl.

On old rails and trunks.  On old rails near Hintonburg, April 18th, 1896 ; abundant on the upper part of a fallen hemlock near McKay's Lake, Beechwood, Sept. 29th, 1896; also on old rails at Aylmer, Que.

### 301.  Physcia hispida, (Schreb.) Tuckerm.

On trees, but generally on boulders with us.  On black ash trees in a swamp in Stewart's Bush along the C.A.R., April 12th, 1895 ; on boulders in a pasture along Brigham's Creek, near Leamy's Lake, May 7th, 1897.

### 302.  Physcia obscura, (Ehrh.) Nyl.

Trunks, dead wood and rocks.  On the bark of white cedar in Stewart's Bush, April 12th, 1895 ; on granite boulders, Rockcliffe Park ; on trees and limestone and granite boulders at Britannia ; common on trees and rocks at Hull and Aylmer, Que. ; on trunks in Dow's Swamp ; and on stones in Ottawa East : on old fence boards at Billing's Bridge ; on trunks and boulders between Brigham's Creek and Leamy's

Lake, Hull, Que. ; on old boards at Stittsville ; on rocks, trunks and old rails, King's Mountain, Chelsea, Que., May 22nd, 1897.

303. **Physcia setosa**, (Ach.) Nyl.

On rocks, and upon mosses, and trunks. On trunks along the Beaver Meadow, west of Hull, Que, April 26th, 1891 ; on trunks "Pine Hill," Rockcliffe Park ; on black ash trunks, Cowley's Farm, west of Hintonburg ; on black ash trunks amongst moss in woods north of Beechwood Cemetery ; on beech trees in woods one mile south-east of Billing's Bridge, on various trees in woods at Leamy's Lake ; on trees in a swamp at Stittsville, May 14th, 1897.

304. **Physcia adglutinata**, (Flœrk.) Nyl.

On trees and shrubs. On beech trunks at Beechwood Cemetery, April 23rd, 1892.

### IX. PYXINE, Fr.

305. **Pyxine sorediata**, Fr.

On trunks in woods. On black ash in Stewart's Bush, April 12th, 1895 ; on beech trunks, "Pine Hill," Rockcliffe Park ; in woods north of Beechwood Cemetery ; on ash, balsam fir and other trees in woods west of Beaver Meadow, Hull, Que.; on trees at Britannia ; on trunks in woods at Leamy's Lake, May 7th, 1897.

### X. UMBILICARIA, Hoffm.

306. **Umbilicaria Muhlenbergia**, (Ach.) Tuckerm.

On perpendicular rocks near the summit of King's Mountain, west of Chelsea, Que., Sept., 1884 ; also May 22nd, 1897.

307. **Umbilicaria vellea,** (L.) Nyl.

On the face of a perpendicular rock near the summit of King's Mountain west of Chelsea, May 22nd, 1897.

308. **Umbilicaria Dillenii**, Tuckerm.

On the faces of perpendicular rocks near the summit of King's Mountain west of Chelsea, May 22nd, 1897.

### XI. STICTA, (Schreb.) Fr.

309. **Sticta amplissima**, (Scop.) Mass.

On large trunks in old woods, not rare. Common in Rockcliffe Park and McKay's woods, April 16th, 1891 ; in Dow's Swamp ; on rear of Skead's Farm, Richmond road ; also on trees at Carleton Place ; on trees in the swamp west of Hull Station ; on basswood trunks in woods near Hull Cemetery ; on trunks King's Mountain and near Chelsea, May 22nd, 1897.

310. **Sticta pulmonaria,** (L.) Ach.

On large old trees in thick woods, common. Common in Rockcliffe Park and McKay's woods and woods north of Beechwood Cemetery, April 20th, 1891 ; on trees at Carleton Place ; on trees in a swamp at Stittsville ; on trunks King's Mountain and near Chelsea, May 22nd, 1897 ; on trees in a swamp west of Hull Station.

### XII. NEPHROMA, Ach.

311. **Nephroma Helveticum,** Ach.

On rocks at King's Mountain, west of Chelsea, Sep., 1884 ; also on rocks below the summit, May 22nd, 1897.

312. **Nephroma lævigatum,** Ach.

On large boulders in old woods. In woods north of the Aylmer road and west of Hull, Que., April 27th, 1895 ; in McKay's woods and in Beechwood Cemetery ; on rocks near the summit of King's Mountain, May 22nd, 1897.

313. **Nephroma parile,** Nyl.

On rocks at King's Mere, west of Chelsea, Que., Sept 3rd, 1884.

### XIII. PELTIGERA, (Willd.) Fee.

314 **Peltigera venosa,** (L.) Hoffm.

On earth along the broken bank of the Lievre River at Buckingham, Que., May 14th. 1896.

315. **Peltigera aphthosa,** (L.) Hoffm.

On rocks, logs and earth and among mosses in swampy woods. On earth and old logs in wet woods east of the Beaver Meadow, west of Hull, Que., April 26th, 1891 ; on damp rocks by the Lievre River, Buckingham, Que.; on dead logs at Carleton Place ; on old logs in a swamp at Stittsville, May 14th, 1897.

316. **Peltigera horizontalis,** (L.) Hoffm.

On moist rocks amongst mosses. In woods near the lake at the head of the Beaver Meadow west of Hull, Que., May 16th, 1896.

317. **Peltigera rufescens,** (Neck ) Hoffm.

On earth, rocks, the bases of trees and amongst moss. On rocks in Rockcliffe Park, Beechwood Cemetery, and Ottawa East on a boulder ; on rocks rear of Cowley's Farm west of Hintonburg ; on old logs Dow's Swamp ; on earth at Britannia ; common on earth and rocks south of the Aylmer road, Hull, Que.; on earth in a swamp at Stittsville ; on rocks near summit of King's Mountain, May 22nd 1897.

318. **Peltigera canina**, (L.) Hoffm.

On earth, rocks and the bases of trees in cool woods. On earth in pine woods Rockcliffe Park ; on earth Ottawa East ; on earth and rocks in woods west of the Beaver Meadow, Hull, Que. ; on old logs in woods Carleton Place ; on the earth at the base of trees in woods, Leamy's Lake ; on earth in a swamp at Stittsville ; very common on earth, old wood and rocks, King's Mountain, May 22nd, 1897.

## XIV. SOLORINA, Ach.

319. **Solorina saccata**, (L.) Ach.

On calcareous earth in the damp crevices of the limestone ledges facing the Ottawa below Governor's Bay, Rockcliffe Park, April 17th, 1895.

## XV. PANNARIA, Delis.

320. **Pannaria lanuginosa**, (Ach.) Kœrb.

On limestone ledges along the cliffs of Rockcliffe Park, April 17th 1895 ; on overhanging rocks along the Beaver Meadow west of Hull, Que. ; also along the Ottawa River on limestone cliffs near Tetreauville, Little Chaudiere ; on limestone rocks Ottawa East ; also near the Experimental Farm ; very common on the faces of damp rocks King's Mountain, May 22nd, 1897.

321. **Pannaria leucosticta**, Tuckerm.

On trunks and rocks, rare. On bark of, balsam poplar in woods south of the Aylmer road, west of Hull, Que., April 27th 1895 ; on boulders Rockcliffe Park ; and on stones West End Park ; on beech trunks in woods one mile south east of Billings Bridge ; on beech trunks north or Beechwood Cemetery ; on rocks King's Mountain, May 22nd, 1897.

322. **Pannaria microphylla**, (Schm) Delis.

Forming a thick crust on rocks. On boulders in woods north of Aylmer Road, west of Hull, Que., April 27th, 1895 ; on boulders, "Pine Hill," Rockcliffe Park and in Beechwood Cemetery ; quite common on damp boulders, King's Mountain, May 22nd, 1897.

323. **Pannaria lepidiota**, Fr.

On earth and amongst moss on rocks. On moss on a stone in woods, south of Aylmer Road, west of Hull, Que., April 27th, 1895.

324. **Pannaria nigra**, (Huds.) Nyl.

On limestone rocks by the cliffs along the Ottawa, Rockcliffe Park, April 16th, 1891, ; on limestone rocks in a field by the Beaver Meadow

Creek, west of Hull, Que. ; on limestone rocks at Leamy's Lake ; on limestone rocks between Aylmer and King's Mountain, May 22nd, 1897.

### XVI. COLLEMA, Hoffm.

325. **Collema myriococcum**, Ach.

Growing on moss, on limestone rocks by the Ottawa, below Governor's Bay, Rockcliffe Park, April 16th, 1891.

326. **Collema pulposum**, (Bernh.) Nyl.

On earth on limestone rocks " Pine Hill," Rockcliffe Park, April 16th, 1896.

327. **Collema crispum**, Borr.

On calcareous earth in the cutting for the Aylmer Railway west of Hull, Que., April 27th, 1895 ; on earth in crevices of rocks at King's Mountain near Chelsea, Que., May 22nd, 1897.

328. **Collema limosum**, Ach.

On calcareous earth in the cutting for the Aylmer Railway west of Hull, Que.; very rare.   April 27th, 1895.

329. **Collema floculosa**, Nyl.

On limestone rocks below Governor's Bay, Rockcliffe Park, April 17th, 1895 ; on naked limestone rocks in woods south of the Aylmer Electric Railway, west of Hull, Que. ; very rare, May 16th, 1895.

330. **Collema tenax**, (Ach.) Tuckerm.

On calcareous earth on wet rocks in the cutting for the Aylmer Electric Railway west of Hull, Que., April 27th, 1895 ; on calcareous earth along the limestone ledges at Rockcliffe Park, April 12th, 1896.

### XVII. LEPTOGIUM, Fr.

331. **Leptogium tenuissimum**, (Dicks.) Koerb.

On sandy earth, on old fence rails along the Richmond Road west of Hintonburg, April 18th, 1896.

332. **Leptogium lacerum**, (Ach.)

On limestone rocks amongst moss in the cutting for the Aylmer Electric Railway, west of Hull, Que., April 27th, 1895 ; on limestone rocks by the Ottawa below Governor's Bay, Rockcliffe Park ; on damp rocks, King's Mountain, May 22nd, 1897.

333. **Leptogium pulchellum**, (Ach.) Nyl.

On trees in woods, Rockcliffe Park, Sep. 16th, 1889, very rare.

**334. Leptogium Tremelloides** (L.) Fr.

On rocks and trunks ; common. On boulders and trees north of the Aylmer Road, west of Hull, Que., April 27th, 1895 ; on trees in Dow's Swamp ; on boulders in Beechwood Cemetery ; in McKay's woods, and on " Pine Hill," Rockcliffe Park ; on old logs and rocks King's Mountain, May 22nd, 1897.

**335. Leptogium chloromelum,** (Sw.) Nyl.

On old rails near Aylmer and on damp rocks near the summit of King's Mountain, west of Chelsea, May 22nd 1897.

**336. Leptogium respulinum,** Ach.

On limestone rocks near the Ottawa below Governor's Bay, Rockcliffe Park, April 16th, 1891.

XVIII. PLACODIUM (DC.)

**337. Placodium elegans,** (Link.) DC.

On a large boulder in woods, Governor's Bay, Rockcliffe Park, April 16th, 1895.

**338. Placodium aurantiacum,** (Lightf.)

On trees and rocks ; also on dead wood. On a granite boulder in woods, Governor's Bay, Rockcliffe Park, April 17th, 1895 ; on limestone rocks at Britannia ; on boulders in woods west of Hull and on boulders at Leamy's Lake, Oct. 9th, 1896.

**339. Placodium cerinum,** (Hedw.)

Common on trees, on dead wood and mosses. On maple and poplar bark Stewart's bush near the C. A. R. track, April 12th, 1895 ; on dead trees and living ash bark at Britannia ; on poplar bark south of the Aylmer Electric Road, west of Hull, Que. ; on black ash Ottawa East ; on trunks in woods at Leamy's Lake ; on poplar trees in a swamp at Stittsville ; on old rails and trunks King's Mountain, May 22nd, 1897.

**340. Placodium vitellinum,** (Ehrh.)

On dead wood and rocks. On old pine rails at Britannia, April 20th, 1895 ; on cedar rails Ottawa East ; on boulders in pasture by Brigham's Creek - on old rails and logs, King's Mountain, May 22nd, 1897.

**341. Placodium vitellinum,** (Ehrh.) var. **aurellum,** Ach.

On granite boulders in woods, Governor's Bay, Rockcliffe Park, April 17th, 1895 ; on boulders in woods north of the Aylmer road west of Hull, Que.; on boulders Ottawa East ; on boulders in pastures by Brigham's Creek, May 7th, 1897.

### XIX. LECANORA, Ach.

**342. Lecanora muralis,** (Schreb.) var. **saxicola.** Schaer.

Very common on both gaanite and limestone boulders, Governor's Bay, Rockcliffe Park, April 17th, 1895 ; quite common on boulders south of the Aylmer road and west of Brigham's Creek, Hull, Que.; on rocks between Chelsea and King's Mountain, May 22nd, 1897.

**343. Lecanora pallida,** (Schreb.) Schaer.

On young pine trees Carleton Place, May 7th, 1892 ; on a pine trunk Rockcliffe Park ; on trunks in woods west of Hull station, also in woods near Leamy's Lake ; on trunks in a swamp at Stittsville ; on rails and trunks, King's Mountain, May 22nd, 1897.

**344. Lecanora pallida,** (Schreb.) var. **cancriformis,** Tuck.

On living pine trees in woods near the C. P. R. bridge over the Ottawa, west of Hull, April 27th, 1895 ; on beech trees, Rockcliffe Park.

**345. Lecanora subfusca,** (L.) var. **allophana,** Ach.

On living pine trees in woods near the C. P. R. bridge, over the Ottawa, Hull, Que., April 27th, 1895 ; on old cedar rails in McKay's woods and Ottawa East ; on maple and beech trunks Rockcliffe Park and Beechwood ; on maple trunks one mile south east of Billings Bridge ; on basswood bark in Dow's Swamp; on trees in a swamp west of Hull station and on boulders in a field by the Aylmer Road ; on trunks in woods by Leamy's Lake and on boulders by Brigham's Creek ; on trunks in a swamp at Stittsville ; on trunks, common, King's Mountain, May 22nd, 1897.

**346. Lecanora subfusca,** (Schreb.) var. **coilocarpa,** Ach.

On beech bark in woods, Rockcliffe Park, April 15th, 1891 ; on beech bark in woods one mile south east of Billings Bridge ; on trunks on King's Mountain, May 22nd, 1897.

**347. Lecanora subfusca** var. **argentata,** Ach.

On small trees at the western base of King's Mountain, west of Chelsea, May 22nd, 1897.

**348. Lecanora Hageni,** Ach.

On old rails near McKay's Lake, April 23rd, 1891 ; on cedar bark on fences, Ottawa East ; on old cedar rails along the Richmond Road above Hintonburg, April 18th, 1896.

**349. Lecanora atra,** (Huds.) Ach.

On young beech trees at Buckingham, Que., May 14th, 1896 ; on limestone shingle at Britannia, April 20th, 1895.

350. **Lecanora badia,** (Pers.) Ach.

On limestone rocks in woods, north of the Aylmer Road, west of Hull, Que., April 27th, 1895.

351. **Lecanora varia,** (Ehrh.) Nyl.

On bark of trees old boards and fence rails. On pine bark in woods near the C. P. R. bridge west of Hull, Que., April 27th, 1895.

352. **Lecanora varia,** var. **symmicta,** Ach.

On trees and fences ; not rare. On bark of living pine trees in woods west of the Beaver Meadow, Hull, Que., April 27th, 1895.

353. **Lecanora varia,** var. **sæpincola,** Fr.

On the board fence in the cutting for the Aylmer Electric Railway, west of Hull, Que., April 27th, 1895 ; on old fences at Stittsville ; on old fences between Aylmer and King's Mountain, May 22nd, 1897.

354. **Lecanora pallescens,** (L.) Schaer.

On birch trees near Ottawa 1884 ; on trunks at the base of King's Mountain, May 22nd, 1897.

355. **Lecanora privigna,** var. **pruinosa,** Auct.

On limestone rocks, in woods, south of the Aylmer Road, west of Hull, April 27th, 1895 ; on boulders along the road and in fields, Rockcliffe Park ; on limestone boulders in a pasture by Brigham's Creek ; abundant on rocks, King's Mountain, May 22nd, 1897.

### XX. RINODINA, Mass.

356. **Rinodina Ascociscana,** Tuck.

On beech trees in McKay's woods near the Lake, April 24th 1891 ; on beech trees in woods one mile south-east of Billings Bridge, April 19th, 1897.

357. **Rinodina sophodes,** (Ach.) Nyl.

On bark of young red maples in Stewart's bush south of the C. A. R. track April 12th, 1895 ; very common on bark of young and old red maple trees along the lake at Britannia ; on beech trees in woods one mile south-east of Billings Bridge, April 19th, 1897.

358. **Rinodina constans,** Nyl.

On beech trees in Beechwood Cemetery, April 20th, 1891.

### XXI. PERTUSARIA, DC.

359. **Pertusaria multipunctata,** (Turn.) Nyl.

On old trees in Rockcliffe Park and Beechwood Cemetery, April 27th, 1892 ; on butternut trees along the Aylmer Road west of Hull ; on trunks of the same near the entrance to Hull Roman Catholic Cemetery; on old cedar rails between Aylmer and King's Mountain, May 22nd 1897.

### 360. **Pertusaria communis,** DC.

On bark of old trees.   On maple trees in Stewart's Bush near the C.A.R. track, April 12th, 1898 ; on beech trees in woods one mile south-east of Billings' Bridge ; on old fence rails along the Richmond Road west of Hintonburgh ; on trees in woods west of Hull ; on trees in woods at Leamy's Lake ; on trees and old logs, Aylmer and King's Mountain.

### 361. **Pertusaria velata,** (Turn.)

On an ash tree in a swamp, Britannia, April 20th, 1895 ; on beech trunks, "Pine Hill," Rockcliffe Park ; on old rails near Aylmer and at the base of King's Mountain ; on butternut trees along the Aylmer Road west of Hull.

### 362. **Pertusaria leioplaca** (Ach.)

On beech trees in woods at Buckingham, Que., May 14th, 1896.

### 363. **Pertusaria Wulfenii,** DC.

On beech trees in woods, Rockcliffe Park, April 21st, 1891 ; on the base of beech trees in woods one mile south-east of Billings' Bridge ; on beech trees in woods west of Hull Station ; also on beech trunks near Leamy's Lake.

### XXII. CONOTREMA, Tuckerm.

### 364. **Conotrema urceolatum,** (Ach.) Tuckerm.

On bark of bitter nut hickory in woods north of the Aylmer Road and west of Hull, Oct. 4th, 1884.

### XXIII. GYALECTEA, (Ach.)

### 365. **Gyalectea lutea,** (Dicks.) Tuckerm.

On the bark of hemlock trees in woods north of Beechwood Cemetery, April 27th, 1892.

### XXIV. THELOTREMA, (Ach.)

### 366. **Thelotrema lepadinum,** Ach.

On black ash trees in a swamp west of the entrance to Beechwood Cemetery, Oct. 16th, 1884.

## XXV. STEREOCAULON, Schreb.

### 367. Stereocaulon paschale, (Ach.)

Abundant on damp shingle along the C. P. R. west of Britannia, April 20th, 1895 ; on boulders in pastures, Ottawa East, and almost everywhere around Ottawa ; on boulders around Hull, quite common ; also on boulders near the Catholic Cemetery, Hull ; on boulders in fields at Stittsville ; on boulders at King's Mountain.

## XXVI. CLADONIA, Hoffm.

### 368. Cladonia alcicornis, Flœrk.

On limestone boulders, in shade, south of the Aylmer Electric Railway and west of the C. P. R., Hull, Que., April 27th, 1895 ; on earth on stones along Brigham's Creek ; on limestone rocks between Aylmer and King's Mountain.

### 369. Cladonia mitrula, Tuckerm.

On earth at the base of pine stumps west of Britannia, Oct. 11th, 1890.

### 370. Cladonia cariosa, Flœrk.

On earth on stones by a fence, Ottawa East, April 14th, 1897 ; on earth at the base of a stump in woods, Leamy's Lake ; on old rails near Aylmer.

### 371. Cladonia pyxidata, Fr.

On earth, rocks, old logs and old fence rails. Old rails, Stewart's Bush, April 12th, 1895 ; on boulders, earth and old stumps, Rockcliffe Park ; common on earth and pine stumps at Britannia ; on old pine logs and limestone rocks, Aylmer Road, Hull, Que. ; on boulders, borders of Dow's Swamp ; on boulders, Ottawa East and Billings' Bridge ; on boulders along Brigham's Creek ; on old rails and stones near Aylmer.

### 372. Cladonia fimbriata, (L.) Fr.

On pine stumps, "Pine Hill," Rockcliffe Park, April 14th, 1895 ; on cedar stumps in Dow's Swamp ; and also in a swamp east of Beech-wood Cemetery ; on the base of stumps in woods west of Hull Station ; on the base of stumps in a swamp at Stittsville ; on old rails near Aylmer and King's Mountain.

### 373. Cladonia fimbriata var. tubæformis, Fr.

On pine stumps and old pine logs at Rockcliffe Park, April 17th, 1895 ; on pine logs and stumps at Britannia ; on rotten pine logs and stumps south of the Aylmer Road, west of Hull, Que. ; on old logs in Dow's Swamp ; on dead wood in woods near Leamy's Lake ; on old logs in a swamp at Stittsville ; on old rails near Aylmer and King's Mountain.

374. **Cladonia gracilis**, var. **verticillata**, Flœrk.

Quite common on earth at Britannia, along the Can. Pac. Railway west of the station, April 20th, 1895 ; on boulders in woods, Rockcliffe Park ; on earth in woods at Leamy's Lake.

375. **Cladonia gracilis**, var. **hybrida**, Schær.

On pine stumps and earth ; common. Pine stumps, Rockcliffe Park, April 17th, 1895 ; on earth and pine stumps at Britannia ; on old pine logs and stumps at Stittsville and Carleton Place, and on the same at Aylmer ; on earth in woods, Leamy's Lake ; on old rails and logs near Aylmer.

376. **Cladonia gracilis**, var. **elongata**, Fr

On old pine logs in a swamp at Stittsville, May 14th, 1897.

377. **Cladonia squamosa**, Hoffm.

On old pine stumps at Britannia, April 20th, 1895 ; on the base of a stump in woods, Leamy's Lake.

378. **Cladonia furcata**, var. **crispata**, Flœrk.

A small clump on a pine stump a mile west of Britannia, April 20th, 1895 ; on old pine logs south of Aylmer Road, west of Hull ; on damp earth in woods along the cliff, Rockcliffe Park, May 7th, 1896.

379. **Cladonia furcata**, var. **racemosa**, Flœrk.

On old logs in woods in Rockcliffe Park ; also on pine stumps at Britannia, April 20th, 1895.

380. **Cladonia rangiferina**, (L.) Hoffm.

On pine stumps in Rockcliffe Park, April 17th, 1895 ; on earth and pine stumps at Britannia ; on old logs and stumps in a swamp at Stittsville ; on old stumps at Carleton Place ; on old logs and stumps in a swamp east of Beechwood Cemetery ; on old pine stumps and logs, Aylmer Road, west of Hull ; on old logs and earth, King's Mountain.

381. **Cladonia rangiferina**, var. **alpestris**, L.

On rotten pine stumps at Britannia, April 20th, 1895.

382. **Cladonia uncialis**, (L.) Fr.

On rocks on the island at Gilmour's Mill, Chelsea, Que., May 15th, 1896 ; on the summit of King's Mountain.

383. **Cladonia delicata**, (Ehrh.) Flœrk.

On rotten pine stumps at Britannia, April 20th, 1895 ; not uncommon on old pine logs in woods close to the Can. Pac. Railway bridge west of Hull ; on the base of stumps in woods at Buckingham, Que. ; on an old pine log in a swamp at Stittsville.

384. **Cladonia deformis**, (L.) Hoffm.

On a pine stump at Britannia ; very rare. April 20th, 1895 ; on an old stump in a swamp at Stittsville ; in a swamp near Lake Flora, Hull, Que. ; on earth slopes of King's Mountain.

385. **Cladonia digitata,** (L.) Hoffm.

On an old pine stump about a mile west of Britannia, April 20th, 1895 ; on old pine logs south of the Aylmer Road, west of Hull.

386. **Cladonia cristatella,** Tuckerm.

On cedar rails and pine stumps and old logs in Stewart's Bush, April 12th, 1895 ; on old pine stumps, Rockcliffe Park ; very common on pine stumps at Britannia ; on dead pine logs and stumps at Aylmer, Que. ; on logs in a swamp east of Belleville ; on stumps and old logs at Stittsville ; on logs and stones in woods near Leamy's Lake ; on old stumps and pine logs near Aylmer.

XXVII. BÆOMYCES, (Pers.) DC.

387. **Bæomyces æruginosus,** (Scop.) DC.

On dead pine wood in cool woods. In woods at Meeche's Lake, Que., Sept. 23rd, 1893 ; in woods at Buckingham, Que., May 14th, 1896.

XXVIII. BIATORA, Fr.

388. **Biatora rufo-nigra,** Tuckerm.

On limestone rocks in Rockcliffe Park, April 17th, 1895.

389. **Biatora granulosa,** (Ehrh.) Pœtsch.

On carbonized wood on dead pine stumps one mile above Britannia, April 20th, 1895 ; on burnt logs, King's Mountain.

390. **Biatora rubella,** (Ehrh.) Rabenh.

On bark of maple and ash trees at Stewart's Bush near the Can. Atlantic Railway, April 12th, 1895 ; on ash trees in a swamp at Britannia ; on black ash, white cedar and maple, Aylmer Road, west of Hull ; also on oak bark in Rockcliffe Park ; on trees in a swamp at Stittsville ; on trees in woods at Leamy's Lake, near Hull ; on black ash in a swamp near Hintonburgh, April 18th, 1896.

391. **Biatora fusco-rubella,** (Hoffm.)

Near the base of black ash trees in Stewart's Bush near Canada Atlantic Railway, April 12th, 1895 ; on balsam poplar bark in woods south of the Aylmer Road west of Hull ; on beech trees in woods at Buckingham, Que., May 14th, 1896 ; on black ash east of Beechwood Cemetery, and west of Hull Station ; on the base of young maples in woods, Rideau Park, April 19th, 1897.

392. **Biatora suffusa,** Fr.

On the base of black ash trees in Stewart's Bush near Canada Atlantic Railway, April 12th, 1895 ; on bark of black ash, Aylmer Road, west of Hull ; on basswood bark, Dow's Swamp ; on beech trees in woods near Leamy's Lake ; on black ash bark in the swamp west of Hull Station, Que., April 24th, 1897.

393. **Biatora Schweinitzii**, Fr.

On spruce, pine and beech trees in woods at Rockcliffe Park, April 17th, 1895 ; on white cedar in Dow's swamp ; on old fence rails along the Richmond Road west of Hintonburgh, April 18th, 1896 ; on cedar bark in a swamp at Stittsville ; on spruce trees in woods west of Hull, Que., May 7th, 1892.

394. **Biatora sanguina-atra**, Fr.

On moss on the base of trees in Dow's Swamp; May 2nd, 1896 ; on earth at the base of trees along the cliff in Rockcliffe Park, April 22nd, 1896 ; on moss in woods west of the Beaver Meadow, Hull, Que., Oct. 20th, 1884.

395. **Biatora varians**, Fr.

On bark of young maples at Casselman ; and at Aylmer, Que., May 5th, 1891 ; on alder bark in Dow's Swamp, May 2nd, 1892.

396. **Biatora oxyspora**, (Tul.)

On *Parmelia Borreri* in McKay's Woods, near the Lake, April 23rd, 1891.

397. **Biatora Laureri**, (Hepp.)

On the bark of dead and living beech trees in woods near McKay's Lake ; on beech trunks, "Pine Hill," Rockcliffe Park, April 16th, 1896.

398. **Biatora sphæroides**, (Dicks.)

On roots of trees at Ottawa, 1884 ; on the bases of trees in woods at Carleton Place, May 12th, 1892 ; on moss on rock at Rockcliffe Park, May 7th, 1896,

399. **Biatora hypnophiba**, Turn.

On moss on rocks or rails. On moss on damp rocks, Rockcliffe Park, April 16th, 1891 ; also on moss on an old log in Beechwood Cemetery, April 14th, 1896.

400. **Biatora Macounii**, Eckfeldt. (N. sp.)

On granite boulders in woods at Rockcliffe Park, April 17th, 1895 ; also on boulders in woods south of the Aylmer Road, Hull, Que., April 27th, 1895.

XXIX. HETEROTHECIUM, Flot.

401. **Heterothecium pezizoideum**, (Ach.) Flot.

On moss on the base of a tree at Carleton Place, Oct. 21st, 1891.

XXX. BUELLIA, De Not.

402. **Buellia parasema**, Ach.

Not uncommon on the bark of growing pine trees. In McKay's Woods and "Pine Hill," Rockcliffe Park, April 20th, 1891 ; on young pines, along the Ottawa River west of Hull, Que. ; on pines, at King's Mountain, near Chelsea, Que ; on ash trees in a swamp north of Beechwood Cemetery, April 27th, 1892 ; on trees in a swamp at Stittsville, May 14th, 1897.

### 403. Buellia myriocarpa, (DC.)

On old fence rails ; doubtless common.  On stones in fields near Britannia, April 20th, 1895 ; on old fence rails in McKay's Woods, quite common. April 23rd, 1891 ; on old rails at Stittsville, May 14th, 1897 ; also at Buckingham, Que ; on old fence rails near Hintonburgh, April 18th, 1896.

### 404. Buellia papillata, (Sommerf.) Tuck.

On moss on old fence rails at Carleton Place, Oct. 11th, 1889.

### 405. Buellia Pertusaricola, Willey.

On the bark of aspen poplar, but parsitic on *Pertusaria communis* n woods by the Beaver Meadow near Hull, Que., Oct. 16th, 1889.

### XXXI. GRAPHIS, Ach.

### 406. Graphis scripta, (Ach.)

Very commom on trunks of all kinds in woods around Ottawa.  On black cherry, beech, maple and oak bark at Aylmer, May 6th, 1891 ; on butternut, birch and beech at Hull, Que., April 28th, 1891 ; on maple, basswood and ironwood in Beechwood Cemetery, April 26th, 1892 ; on balsam fir at Stittsville, May 14th, 1897 ; on blue beech at Leamy's Lake, Hull, Que. ; also abundant on trees at King's Mountain, Chelsea, Que., May 22nd, 1897 ; on alder trunks in Dow's Swamp ; also in Rideau Park, April 19th, 1897.

### 407. Graphis recta, Humb.

Not uncommon on the bark of yellow and canoe birch in woods. In woods along the Beaver Meadow, Hull, Que., May 16th, 1896 ; also on the same at Buckingham, Que., May 14th, 1896.

### XXXII. OPEGRAPHA, Humboldt.

### 408. Opegrapha varia, Pers.

On butternut bark in woods along the Aylmer Road west of Hull, Que., April 23rd, 1891 ; also on cedar bark at King's Mountain, near Chelsea, Que., May 22nd, 1897 ; on cedar bark in Dow's Swamp, May 2nd, 1897.

### XXXIII. ARTHONIA, Ach.

### 409. Arthonia astroidea, Ach.

On bark, quite common in woods around Ottawa.  On bark of *Juglans cinerea* in woods along the Aylmer Road west of Hull, Que., April 28th, 1891 ; on *Abies balsamea* and young pines in Dow's Swamp, April 23rd, 1892 ; in woods near Aylmer, Que.

### 410. Arthonia Swartziana, Ach.

Not uncommon on oak and ironwood trees near Aylmer, Que., May 6th, 1891 ; on maple trees at King's Mountain, near Chelsea, Que., May 22nd, 1897.

### 411. Arthonia lecideella, Nyl.

On various young trees and shrubs.  On *Acer spicatum* at Aylmer, Que., May 6th, 1891 ; quite common on young *Acer rubrum* in Stewart's Bush and Rideau Park ; on young maple trees at Buckingham, Que., May 14th, 1896 ; on young red maples near Hintonburgh, April 18th, 1896.

### 412. Arthonia spectabilis, Flot.

On thick bark of old trees ; common.  on basswood and sugar maple bark at Carleton place, May 7th, 1882 ; on bark of *Carya amara* in woods west of Hull, Que., September 21st, 1889 ; on old maples at Casselman ; on *Juglans cinerea* at Aylmer, Que., May 6th, 1891 ; on maple trees in Rideau Park, near Billings' Bridge, April 19th, 1897.

### 413. Arthonia tædiosa, Nyl.

On young beech and maples, at Buckingham, Que., May 14th, 1896.

### 414. Arthonia dispersa, (Schrad.) Nyl.

On bark of young sugar maples at Ottawa.  Collected in Beechwood Cemetery, April 16th, 1892.

### XXXIV. MYCOPORUM, (Flot.) Nyl.

### 415. Mycoporum pycnocarpum, Nyl.

On oak bark in woods by the lake near Aylmer, May 6th, 1891 ; also on bark of young *Acer rubum*, at Britannia, April 20th, 1895.

### XXXV. CONIOCYBE, Ach.

### 416. Coniocybe furfuracea, (L.) Ach.

On the roots of trees in woods.  On earth on pine roots in woods north of Ironsides, Que., Sept. 16th, 1891 ; on roots of trees in Dow's Swamp, Oct. 12th, 1887.

### XXXVI. CALICIUM, Pers.

### 417. Calicium subtile, Fr.

On dead cedar stump in Dow's Swamp, Sept. 26th, 1891.

### XXXVII. ENDOCARPON, Hedw.

### 418. Endocarpum, fluviatite, DC.

On stones in the Beaver Meadow Brook below the C. P. Ry. bridge near Hull Station, Que., April 24th, 1897 ; on rocks in a brook at Meeche's Lake, near Chelsea, Que., Sept. 23rd, 1893 ; on limestone rocks, close to the Ottawa River, Gatineau Ferry, Rockcliffe, Nov. 12th, 1896.

## XXXVIII. TRYPETHELIUM, Spreng.

### 419. Trypethelium virens, Tuck.

On beech trunks on "Pine Hill" in Rockcliffe Park, and Beech-wood Cemetery, April 26th, 1891 ; on beech trees at Casselman, and at King's Mountain near Chelsea, Que., May 22nd, 1897.

## XXXIX. PYRENULA, (Ach.)

### 420. Pyrenula punctiformis, (Ach.)

On bark of trees at Carleton Place, May 12th, 1892 ; on maple trees near Hintonburgh, April 18th, 1896 ; on trunks of sugar maple, " Pine Hill," Rockcliffe Park, April 16th, 1896.

### 421. Pyrenula gemmata, (Ach.)

On old maple trunks in McKay's Woods near the lake, April 16th, 1891.

### 422. Pyrenula mamillana, (Ach.)

On bark of maple trees in old woods, Carleton Place, May 12th, 1892.

### 423. Pyrenula nitida, Ach.

Quite common on beech trunks around Ottawa. On beech trees in Rockcliffe Park and McKay's Woods, April 26th, 1891 ; on beech trees at Moose Creek and Casselman, Sept. 6th, 1891 ; on trees at Aylmer, Que., May 6th, 1891.

### 424. Pyrenula thelena, Ach.

On canoe birch in woods along the Beaver Meadow, Hull, Que. April 24th, 1897.

### 425. Pyrenula fallaciea.

On bark of young maples at Chelsea, Que., May 15th, 1891.

### 426. Pyrenula cinerella

On young maple trees at Ottawa, May 7th, 1892.